A First Course in

Atmospheric
Numerical Modeling

A First Course in

Atmospheric Numerical Modeling

Alex J. DeCaria

Glenn E. Van Knowe

Sundog Publishing
Madison, Wisconsin

Ordering information: Instructors and students should visit www.sundogpublishing.com/ordering/ to take advantage of discounts available to students who order directly from the publisher. Free examination copies may be requested by qualifying instructors.

Online resources: If available, electronic materials relevant to courses utilizing this book will be posted on the publisher website at www.sundogpublishing.com.

Sundog Publishing, LLC, Madison, Wisconsin

DeCaria, Alex J. (Alex Joseph)
    A first course in atmospheric numerical modeling /
Alex J. DeCaria, Glenn E. Van Knowe.
    pages cm
    Includes index.
    LCCN 2013945215
    ISBN 978-0-9729033-4-9 (softcover)
    ISBN 978-0-9729033-6-3 (e-book)

    1. Meteorology--Mathematical models--Textbooks.
2. Atmospheric physics--Mathematical models--Textbooks.
3. Atmospheric models--Textbooks.   I. Van Knowe, Glenn E.
(Glenn Earnest) II. Title.

QC861.3.D43 2014   551.51'01'5118
                          QBI13-1617

10 9 8 7 6 5 4 3 2 1

Cover design by Linda J. Weidemann.

This book is dedicated to our wives, Sandi Van Knowe and Marcia DeCaria, without whose encouragement, efforts, and patience this book would still be just an idea.

Acknowledgments

Alex DeCaria thanks all the students who have taken his modeling course over the past 12 years, for allowing him to hone his teaching skills and refine the material presented in the book. He also thanks Millersville University for supporting the writing of the book through a sabbatical leave.

Glenn Van Knowe thanks the staff at MESO for suggestions and editorial comments on the chapters concerning subgrid-scale processes, data assimilation, model verification and validation, model output postprocessing, and model applications.

We are grateful to Mike Kaplan for reviewing the manuscript and providing constructive feedback. Finally both authors thank their wives, not only for providing encouragement but also for actively proofreading and editing the text.

Contents

CHAPTER 1

Introduction

1.1 Overview

The macroscopic state of the atmosphere is governed by a system of seven equations and seven dependent variables (Chapter 2). These equations are known as the *primitive equations*, or *governing equations*. Several of the equations are partial differential equations containing time derivatives of the dependent variables and are therefore *prognostic* equations. The other equations are *diagnostic*, containing relationships among the dependent variables without time derivatives.

The equations comprise an initial value problem. In theory they can be integrated forward in time to predict the future state of the variables if the initial state of the atmosphere is specified. However, the system of equations contains complex relationships among the variables, and there is no analytical solution that can be written down. Instead, they must be numerically integrated using finite differencing to approximate the time derivatives. The spatial derivatives are either approximated using finite-differences (Chapters 3 and 4) or are spectrally transformed (Chapter 11). Regardless of the technique, numerical solutions are inherently approximate and introduce errors; it is fundamentally impossible to achieve a perfectly accurate solution to the original equations.

Numerical integration requires that the atmosphere be represented as a discrete entity with the values of the variables defined at grid points, even though the actual atmosphere is a physically continuous medium. Many important atmospheric processes occur on spatial scales that are smaller than can be resolved by numerical models, and these subgrid-scale processes must somehow be parameterized in terms of the model variables (Chapter 10).

Because of errors associated with finite differencing, parameterization of subgrid-scale processes, and incomplete knowledge of initial conditions, no numerical model will ever give an exact solution to the initial value problem posed by the governing equations. However, if steps are taken to minimize the errors, the model output can still be useful.

1.2   A Brief History

The primitive equations were already known in the early part of the 20th Century, and Lewis Fry Richardson had even formulated a system for numerically integrating them for weather prediction. Richardson's system was purely an academic exercise. In order to numerically integrate the equations to predict the future state of the atmosphere, a multitude of calculations are required, and computers that could perform these calculations quickly had not yet been invented. Richardson therefore had no way to perform the calculations fast enough to make a forecast that would be available before the predicted weather had actually occurred. Nonetheless, Richardson attempted a numerical integration over a small region of Europe, and compared his results with the actual state of the atmosphere. The forecast was a dismal failure, predicting an unrealistic pressure tendency that was an order of magnitude larger than that observed. He published his method and results in 1922 in the first published book on atmospheric modeling, *Weather Prediction by Numerical Process*.

Although Richardson did not know it at the time, his method was essentially correct but suffered a fatal flaw in that the initial conditions with which he initialized his model were not in hydrostatic nor gradient balance. The imbalances resulted in large-amplitude gravity waves that swamped the meteorologically significant pressure tendency. The problem of formulating balanced initial conditions (Chapter 12) for primitive-equation models was not adequately resolved until the 1960s. Instead, the initialization problem was circumvented by using a modified subset of the primitive equations that do not allow gravity or acoustic waves as solutions. This modified set of equations is based on the quasigeostrophic assumption and is known as the filtered equations (Chapters 6 and 7). The very first operational numerical forecasts were begun in the 1950's and were based on the filtered equations.

Once the initialization problem was resolved in the 1960s, models based on the primitive equations (Chapter 13) gradually supplanted those based on the filtered equations. An additional development was the advent of spectral methods (Chapter 11), which transformed the equations from physical space into spectral space and eliminated the need in many cases to use finite differencing for the spatial derivatives.

1.3   Other Developments

Atmospheric models represent the continuous atmosphere on a discrete grid. Modelers have found that not all of the dependent variables (three wind components, pressure, density, temperature, and humidity) need to be represented at each model grid point, and that the use of staggered grids (Chapter 8) can lead to more accurate and efficient models. In addition, all models have physical as well as numerical boundaries, and how the equations are handled at these boundaries is crucial for model stability and accuracy (Chapter 9).

Numerical integration is an iterative process, advancing the solution forward by discrete time intervals. The length of these time intervals is determined by the specific technique used, but a fundamental constraint on the length of the time interval is the *Courant-Friedrichs-Lewy* or *CFL* condition (Chapter 4). The CFL condition states that for many applications the maximum time interval that can be used is directly proportional to both the model grid spacing and the speed of the fastest wave supported by the model equations. The fastest waves in the atmosphere are acoustic waves and barotropic gravity waves, both of which are excluded by the filtered equations. However, the primitive equations support these fast waves. The constraint on the maximum time interval can be loosened using the semi-implicit and semi-Lagrangian techniques

(Chapter 14).

In Chapter 15 we discuss finite-volume methods which are ideally suited for processes that follow conservation laws. Finite-volume methods are not only useful for simulations of atmospheric dynamics, but are also widely used for modeling the transport of atmospheric constituents such as trace gases and pollutants. Coupling of finite-volume techniques with numerical methods for solving chemical reaction-rate equations leads to the discussion of chemical transport models (CTMs) in Chapter 16.

The results from numerical models must be verified and validated (17) with the goal of providing the most accurate and realistic simulations possible. Postprocessing of model output (18) can enhance its accuracy and usefulness through such methods as model-output statistics (MOS) and ensemble methods.

Finally, the uses of atmospheric numerical models have expanded well beyond the initial task of numerical weather prediction. There are many specialized applications of models across a broad spectrum of industries and research areas. Some of these applications are discussed in Chapter 19.

1.4   Why Another Modeling Book?

Since L. F. Richardson first published *Weather Prediction by Numerical Process* in 1922 there have been published dozens of books discussing atmospheric numerical modeling. However, the books currently available all tend to be too advanced for a first course in modeling, either at the undergraduate or graduate level. Many of these books are fine references, and may even be suitable for advanced courses, but are not suited to the needs of a student or practitioner who is being exposed to the concept of numerical modeling for the first time. Although our book also contains more advanced material than would be expected to be covered in a single, introductory course, it is not as exhaustive nor advanced as many of the existing texts, and is hopefully more approachable. In addition, we have attempted to include a variety of exercises, both theoretical and practical, including programming exercises, since the best way to learn modeling is by actually writing simple models.

## CHAPTER 2

Governing Equations and Assumptions

2.1 Governing Equations

The macrosocpic state of the atmosphere is described by the three components of the wind velocity (u, v, and w), as well as temperature (T), pressure (p), density (ρ), and specific humidity (q). These variables are *time-averaged* or *Reynolds-averaged*, meaning that they are not instantaneous values.[1] These seven variables are related via a set of *governing equations*, also called the *primitive equations*, which consist of three momentum equations (one for each coordinate direction), a mass continuity equation, a water-mass continuity equation, a thermodynamic energy equation, and an equation of state. The governing equations are a closed set of seven equations and seven unknowns. Each equation is briefly described below.

2.1.1 Momentum equations

The three *momentum equations,* also known as the *dynamical equations of motion,* are based on Newton's second law of motion and relate the acceleration of the air in a given direction to the sum of the forces acting in that direction. The forces are the *pressure gradient force,* the *Coriolis force, gravity* (the combination of the gravitational force and

---

[1]Reynolds-averaging is discussed in detail in Chapter 10.

5

the Earth's centrifugal force), and the forces due to turbulent eddies. The three momentum equations in Cartesian coordinates are

$$\frac{Du}{Dt} = -\frac{1}{\rho}\frac{\partial p}{\partial x} + 2\Omega\sin\phi\,v - 2\Omega\cos\phi\,w + F_x$$

$$\frac{Dv}{Dt} = -\frac{1}{\rho}\frac{\partial p}{\partial y} - 2\Omega\sin\phi\,u + F_y \tag{2.1}$$

$$\frac{Dw}{Dt} = -\frac{1}{\rho}\frac{\partial p}{\partial z} + 2\Omega\cos\phi\,u - g + F_z,$$

where Ω is the angular velocity of the Earth and ϕ is latitude. The terms F_x, F_y, and F_z represent the time-averaged fluxes of momentum due to transport by turbulence.

The notation D/Dt refers to the *total derivative* (sometimes called the *material derivative*), which is the derivative following a fluid parcel. The total derivative consists of the local derivative at a fixed point in space, plus a term due to advection, and is expressed mathematically as

$$\frac{D}{Dt} \equiv \frac{\partial}{\partial t} + \vec{V}\cdot\nabla. \tag{2.2}$$

In spherical coordinates, additional terms due to the curvature of the Earth appear in the momentum equations. These curvature terms involve products of the velocity components, as well as latitude and the radius of the Earth. Appendix E discusses Cartesian versus spherical coordinates.

2.1.2   Mass continuity equation

The mass continuity equation expresses the conservation of mass and can be written in two equivalent ways,

$$\frac{\partial\rho}{\partial t} + \nabla\cdot(\rho\vec{V}) = 0 \tag{2.3}$$

or

$$\frac{D\rho}{Dt} + \rho\nabla\cdot\vec{V} = 0. \tag{2.4}$$

Equation (2.3) is known as the *flux form* of the continuity equation. Its physical meaning is that changes in density at a fixed point in space are due to divergence or convergence of mass flux. The second form of the continuity equation, (2.4), relates the change in density following a fluid parcel to the convergence or divergence of the velocity field.

2.1.3 Water-mass continuity equation

The budget of water mass is written as

$$\frac{\partial(\rho q)}{\partial t} + \nabla \cdot (\rho q \vec{V}) = S_O - S_K + F_q \tag{2.5}$$

where S_O and S_K are terms representing sources and sinks of water vapor to the atmosphere, and F_q represents turbulent fluxes of water vapor. Equation (2.5) relates the local change of specific humidity at a fixed point to convergence or divergence of moisture flux. The water-mass continuity equation is very similar to the mass continuity equation, except for the inclusion of the terms for sources and sinks of water vapor.

2.1.4 Thermodynamic energy equation

This equation expresses the first law of thermodynamics, and is

$$c_p \frac{DT}{Dt} - \frac{1}{\rho}\frac{Dp}{Dt} = \dot{Q} + F_T, \tag{2.6}$$

where c_p is the specific heat of air, and $\dot{Q}$ is the rate of diabatic heating due to radiation, sensible, or latent heating. For adiabatic processes, $\dot{Q}$ is equal to zero. The term F_T represents turbulent fluxes of heat.

2.1.5 Equation of state

The three thermodynamic variables, p, T, and ρ, are related through the equation of state. For dry air the ideal gas law,

$$p = \rho R_d T, \tag{2.7}$$

is used. The constant R_d is the specific gas constant for dry air. For moist air, a correction term is included, so that

$$p = \rho R_d T (1 + 0.61q), \tag{2.8}$$

which is often written as

$$p = \rho R_d T_v \tag{2.9}$$

where the virtual temperature, T_v, is defined as $T_v \equiv (1 + 0.61q)T$.

2.1.6   Summary

The terms representing turbulent fluxes of momentum, heat, and moisture are negligible under certain circumstances and can often be omitted. If they are included, then they must be parameterized in terms of the time-averaged variables, a procedure known as *turbulent closure*. We will return to these terms in Chapter 10, but for now we omit them. The two Coriolis terms that involve $\cos\phi$ can also be omitted under many circumstances. Omitting them, and defining the Coriolis parameter $f \equiv 2\Omega\sin\phi$, allows the governing equations (primitive equations) in Cartesian coordinates to be written as

$$\frac{\partial u}{\partial t} + \vec{V}\cdot\nabla u = -\frac{1}{\rho}\frac{\partial p}{\partial x} + fv \tag{2.10}$$

$$\frac{\partial v}{\partial t} + \vec{V}\cdot\nabla v = -\frac{1}{\rho}\frac{\partial p}{\partial y} - fu \tag{2.11}$$

$$\frac{\partial w}{\partial t} + \vec{V}\cdot\nabla w = -\frac{1}{\rho}\frac{\partial p}{\partial z} - g \tag{2.12}$$

$$\frac{\partial\rho}{\partial t} + \nabla\cdot(\rho\vec{V}) = 0 \tag{2.13}$$

$$\frac{\partial(\rho q)}{\partial t} + \nabla\cdot(\rho q\vec{V}) = S_O - S_K \tag{2.14}$$

$$c_p\frac{DT}{Dt} - \frac{1}{\rho}\frac{Dp}{Dt} = \dot{Q} \tag{2.15}$$

$$p = \rho R_d T(1 + 0.61q). \tag{2.16}$$

Six of the equations, (2.10) through (2.15), are *prognostic*, since they contain time derivatives of the dependent variables. This allows them to be integrated forward in time to predict the future value of the variables. The last equation, (2.16), is *diagnostic*, meaning that it cannot be integrated with time to predict the future value of a variable.

In theory, given a set of initial conditions, the governing equations could simply be integrated forward to predict the state of the atmosphere at a future time; however, the governing equations are a complex coupled system of partial differential equations that cannot be solved analytically. Instead, they must be solved on a computer using numerical methods.

2.2   Simplifying Assumptions

Simplifying assumptions are needed for dealing with the complexities of modeling the atmosphere. These assumptions speed up the forecast process and increase the efficiency of the numerical algorithm used to solve the equations. The simplifying assumptions must retain the most important factors that influence the atmosphere. However, they do lead to errors. As computational power increases, the need for simplifying assumptions is lessened, but not eliminated.

2.2.1   Elimination of acoustic waves

The governing equations support many wave-like solutions. Some of these waves represent important processes by which the atmosphere transports energy and momentum, such as Rossby waves and gravity waves. The equations also support waves that are not important for modeling the atmosphere, such as acoustic waves. Inclusion of unimportant waves can sometimes greatly restrict the speed and efficiency of the numerical algorithm used to solve the equations. Elimination of acoustic waves can speed up the forecast process without altering its accuracy.

Acoustic waves can be eliminated by two methods. They can be completely eliminated as solutions by assuming that the atmosphere is completely incompressible, $D\rho/Dt = 0$. In this case the continuity equation becomes

$$\nabla \cdot \vec{V} = 0. \tag{2.17}$$

In order to use the incompressible continuity equation, there are three conditions that must be met:

1. The square of the wind speed must be much less than the square of the speed of sound, $U^2 \ll c_s^2$.

2. The characteristic time scale for changes in density (τ_p) must be much greater than the time scale for changes due to propagation of acoustic waves ($\tau_p \gg L_x/c_s$), where L_x is the characteristic horizontal scale of motion).

3. The vertical scale of the circulation, L_z, must be much less than the scale height of the atmosphere, H_x ($L_z \ll H_x$).

The first two conditions are readily met in the atmosphere. The third is not, since it would only apply to very shallow circulations, such as those confined to the planetary boundary layer.

Allowing density to vary with height, but not allowing it to change locally (so that $\partial \rho / \partial t = 0$) yields another form of the continuity equation,

$$\nabla \cdot (\rho_0 \vec{V}) = 0. \tag{2.18}$$

This form of the continuity equation also eliminates acoustic waves, yet is not restricted to shallow circulations. It is known as the *anelastic continuity equation* and it is widely used in atmospheric numerical models.

2.2.2   The hydrostatic approximation

There are two contributions to the pressure at any given point in the atmosphere. One contribution is from the mass of the atmosphere above the point, called the *hydrostatic pressure*. The other contribution is the *dynamic pressure*, due to the fluid motion. The dynamic pressure is small compared to the hydrostatic pressure for motions whose horizontal scale is much larger than the vertical scale ($L_x \gg L_z$). In these circumstances the only nonzero terms in the vertical momentum equation are the gravity and vertical pressure gradient terms. The result is a diagnostic equation for pressure,

$$\frac{\partial p}{\partial z} = -\rho g, \tag{2.19}$$

known as the *hydrostatic equation*.

Though technically valid only for a fluid at rest, the hydrostatic equation can be used without significant error as long as $L_x \gg L_z$. This condition is met for synoptic and large mesoscale circulations, so most synoptic and many mesoscale numerical models use the hydrostatic equation in place of the full vertical momentum equation.

2.2.3   Boussinesq approximation

For relatively shallow layers of the atmosphere, such as the planetary boundary layer, another simplification to the momentum equations is sometimes possible. This is the *Boussinesq approximation*, in which the density and pressure fields are broken into reference states that vary only with altitude, and departures from the reference state, which may vary both in time and space, so that

$$\rho(x,y,z,t) = \rho_0(z) + \tilde{\rho}(x,y,z,t) \tag{2.20}$$

$$p(x,y,z,t) = p_0(z) + \tilde{p}(x,y,z,t). \tag{2.21}$$

Perturbations of density and pressure are then assumed to be small compared to their reference state values, and the reference state is also assumed to be in hydrostatic balance. With these assumptions the momentum equations in Cartesian coordinates become

$$\frac{\partial u}{\partial t} + \vec{V} \cdot \nabla u = -\frac{1}{\rho_0}\frac{\partial \tilde{p}}{\partial x} + fv \tag{2.22}$$

$$\frac{\partial v}{\partial t} + \vec{V} \cdot \nabla v = -\frac{1}{\rho_0}\frac{\partial \tilde{p}}{\partial y} - fu \tag{2.23}$$

$$\frac{\partial w}{\partial t} + \vec{V} \cdot \nabla w = -\frac{1}{\rho_0}\frac{\partial \tilde{p}}{\partial z} - \frac{\tilde{\rho}}{\rho_0}g. \tag{2.24}$$

Essentially, the Boussinesq approximation substitutes the reference state density, ρ_0, for the full density, ρ, everywhere except in the buoyancy term, which is the term containing gravity.

2.3 Alternate Dependent Variables

2.3.1 Exner function

Equations (2.10) through (2.16) are not the only possible representation for the governing equations. There are choices as to which dependent variables are used to represent the state of the atmosphere. Pressure is often replaced by its dimensionless equivalent, defined as

$$\pi \equiv \left(\frac{p}{P_R}\right)^{R_d/c_p}, \tag{2.25}$$

where P_R is a constant reference pressure, often taken to be 1000 hPa. This dimensionless pressure is named the *Exner function* (after the British meteorologist Felix Exner). Using π in place of p results in the following form of the momentum equations (see Exercise 2.2),

$$\frac{\partial u}{\partial t} + \vec{V} \cdot \nabla u = -c_p \theta \frac{\partial \pi}{\partial x} + fv \tag{2.26}$$

$$\frac{\partial v}{\partial t} + \vec{V} \cdot \nabla v = -c_p \theta \frac{\partial \pi}{\partial y} - fu \tag{2.27}$$

$$\frac{\partial w}{\partial t} + \vec{V} \cdot \nabla w = -c_p \theta \frac{\partial \pi}{\partial z} + g \tag{2.28}$$

where θ is potential temperature. These equations have two advantages over using pressure as a variable. One minor advantage is that density does not appear as a dependent variable. The second, and more significant advantage is that vertical gradients of π are much smaller than vertical gradients of p. This results in a more accurate representation of the vertical pressure gradient term in a numerical model when the derivative is approximated as a finite difference.

2.3.2   Streamfunction and velocity potential

Every two-dimensional fluid flow consists of a rotational flow superimposed on a divergent flow.[2] The rotational (also called *solenoidal*) component of the fluid flow is defined in terms of a scalar called the *streamfunction*, ψ. The solenoidal velocity components of the flow are given in terms of derivatives of the streamfunction,

$$u_s = -\frac{\partial \psi}{\partial y} \tag{2.29}$$

$$v_s = +\frac{\partial \psi}{\partial x}. \tag{2.30}$$

Vorticity expressed in terms of the streamfunction is

$$\zeta = \nabla^2 \psi. \tag{2.31}$$

The divergent (also called the *irrotational*) part of the flow is expressed in terms of a scalar called the *velocity potential*, χ. The components of the irrotational flow are given in terms of derivatives of the velocity potential,

$$u_i = -\frac{\partial \chi}{\partial x} \tag{2.32}$$

$$v_i = -\frac{\partial \chi}{\partial y}, \tag{2.33}$$

---

[2]This is actually true of any 2D vector field.

and the divergence is

$$\delta = -\nabla^2 \chi. \tag{2.34}$$

In global spectral models, streamfunction and velocity potential are often used as the dependent variables in place of the individual wind components u and v. This is done in order to avoid the ambiguity at the North and South Poles as to how u and v are defined.

2.4   Alternate Vertical Coordinates

Our formulation of the governing equations has so far used x, y, and z as the spatial coordinates. It is also possible to use pressure, or some function of pressure, in place of z as a vertical coordinate in numerical models. Some of the more common vertical coordinates are pressure coordinates and sigma-pressure coordinates.

When using an alternate vertical coordinate, arbitrarily denoted here by s, the governing equations must be rewritten in terms of the new coordinate. The vertical velocity in the new coordinate system is defined as

$$\dot{s} = \frac{Ds}{Dt} \tag{2.35}$$

pronounced simply as *s-dot*. To convert vertical velocities from height coordinates to the new coordinate system, the chain rule is used,

$$\frac{Dz}{Dt} = \frac{Dx}{Dt}\left(\frac{\partial z}{\partial x}\right)_{y,s,t} + \frac{Dy}{Dt}\left(\frac{\partial z}{\partial y}\right)_{x,s,t} + \frac{Ds}{Dt}\left(\frac{\partial z}{\partial s}\right)_{x,y,t} + \left(\frac{\partial z}{\partial t}\right)_{x,y,s},$$

or

$$w = u\left(\frac{\partial z}{\partial x}\right)_{y,s,t} + v\left(\frac{\partial z}{\partial y}\right)_{x,s,t} + \dot{s}\left(\frac{\partial z}{\partial s}\right)_{x,y,t} + \left(\frac{\partial z}{\partial t}\right)_{x,y,s}. \tag{2.36}$$

In (2.36) the subscripts on the partial derivatives indicate which independent variables are being held constant. In general, surfaces of the new coordinate are not necessarily parallel to constant height surfaces. This is accounted for in the terms

$$\left(\frac{\partial z}{\partial x}\right)_{y,s,t} \quad \text{and} \quad \left(\frac{\partial z}{\partial y}\right)_{x,s,t},$$

which represent the slope of a surface of constant s. The term

$$\left(\frac{\partial z}{\partial t}\right)_{x,y,s}$$

represents the vertical motion of a coordinate surface itself, which is not necessarily fixed to a particular height.

In the new coordinate system the total derivative is now

$$\frac{D}{Dt} = \frac{\partial}{\partial t} + u\frac{\partial}{\partial x} + v\frac{\partial}{\partial y} + \dot{s}\frac{\partial}{\partial s}. \qquad (2.37)$$

Any vertical derivatives must be put in terms of $\partial/\partial s$ in place of $\partial/\partial z$, which is accomplished using the chain rule,

$$\frac{\partial}{\partial s} = \frac{\partial z}{\partial s}\frac{\partial}{\partial z}. \qquad (2.38)$$

Finally, any horizontal derivatives must be converted into derivatives taken at a fixed s. This relationship is established by

$$\left(\frac{\partial}{\partial x}\right)_z = \left(\frac{\partial}{\partial x}\right)_s - \left(\frac{\partial z}{\partial x}\right)_s \left(\frac{\partial}{\partial z}\right)_x. \qquad (2.39)$$

2.4.1 Pressure coordinates

Under hydrostatic balance, pressure and altitude are related to each other through the hydrostatic equation,

$$\frac{\partial p}{\partial z} = -\rho g. \qquad (2.40)$$

In height coordinates the vertical velocity is defined as

$$w \equiv \frac{Dz}{Dt}, \qquad (2.41)$$

while in pressure coordinates vertical velocity is defined as

$$\omega \equiv \frac{Dp}{Dt}, \qquad (2.42)$$

(often simply referred to as *omega*) and has units of Pascals per second. Since pressure increases downward, a positive omega indicates downward motion.

The relationship between the vertical velocities in pressure and height coordinates is established using (2.36),

$$w = u \left(\frac{\partial z}{\partial x} \right)_{y,p,t} + v \left(\frac{\partial z}{\partial y} \right)_{x,p,t} - \frac{\omega}{\rho g} + \left(\frac{\partial z}{\partial t} \right)_{x,y,p}.$$

The terms

$$\left(\frac{\partial z}{\partial x} \right)_{y,p,t} \text{ and } \left(\frac{\partial z}{\partial y} \right)_{x,p,t}$$

represent the slope of a constant pressure surface. Since this slope is relatively small, these terms are negligible compared to the term $\omega/\rho g$. The term

$$\left(\frac{\partial z}{\partial t} \right)_{x,y,p},$$

which represents the vertical motion of the constant pressure surface itself, is also small compared to the $\omega/\rho g$ term. Therefore, we can use the relation

$$\omega \cong -\rho g w \qquad (2.43)$$

when converting the vertical velocity from pressure to height coordinates or vice versa.

The total derivative in pressure coordinates is

$$\frac{D}{Dt} = \frac{\partial}{\partial t} + u \frac{\partial}{\partial x} + v \frac{\partial}{\partial y} + \omega \frac{\partial}{\partial p}. \qquad (2.44)$$

Vertical derivatives are converted from height to pressure coordinates using (2.38), which in this instance becomes

$$\frac{\partial}{\partial p} = -\frac{1}{\rho g} \frac{\partial}{\partial z}. \qquad (2.45)$$

And finally, any horizontal derivatives are converted into the new coordinate system using (2.39), which becomes

$$\left(\frac{\partial}{\partial x} \right)_z = \left(\frac{\partial}{\partial x} \right)_p - \left(\frac{\partial z}{\partial x} \right)_p \left(\frac{\partial}{\partial z} \right)_x. \qquad (2.46)$$

2.4.2 Sigma coordinates

Coordinate surfaces of height or pressure are horizontal, or nearly so, and therefore can intersect the ground and even be underground, as shown in Fig. 2.1. This is undesirable because it greatly complicates formulating solutions to the model equations at the bottom boundary.

The ideal vertical coordinate would follow the terrain and never intersect it. One way to achieve this is to use the sigma-pressure (or sigma-p) coordinate system. The pressure field is first broken into a reference pressure that only depends on z, and a departure from the reference, $\tilde{p}$, such that

$$p = p_0(z) + \tilde{p}(x, y, z, t). \tag{2.47}$$

We then define the vertical coordinate in terms of the reference pressure,[3]

$$\sigma \equiv \frac{p_0 - p_T}{p_S - p_T} = \frac{p_0 - p_T}{p*}, \tag{2.48}$$

where p_T is the pressure at the model top, and is a constant, while p_S is the value of p_0 at the surface of the model terrain. Note that p_S, and hence $p*$, are functions of x and y. Sigma takes on values between zero and one, with $\sigma = 0$ at the top of the model and $\sigma = 1$ at the model surface. Sigma coordinates are illustrated in Fig. 2.2

Vertical velocity in sigma coordinates is given as

$$\dot{\sigma} = \frac{D\sigma}{Dt}, \tag{2.49}$$

and is simply referred to as *sigma-dot*. It is related to vertical velocity in height coordinates using (2.36), which becomes

$$w = u \left(\frac{\partial z}{\partial x} \right)_{y,\sigma,t} + v \left(\frac{\partial z}{\partial y} \right)_{x,\sigma,t} + \dot{\sigma} \left(\frac{\partial z}{\partial \sigma} \right)_{x,y,t} + \left(\frac{\partial z}{\partial t} \right)_{x,y,\sigma}. \tag{2.50}$$

Vertical derivatives in sigma coordinates are converted from height coordinates using

$$\frac{\partial}{\partial \sigma} = \frac{\partial z}{\partial \sigma} \frac{\partial}{\partial z}. \tag{2.51}$$

---

[3]There are alternate definitions for sigma. One is to define sigma as $\sigma = p/p_S$. There are also sigma coordinate systems defined in terms of height, rather than pressure, which are sometimes used in ocean models.

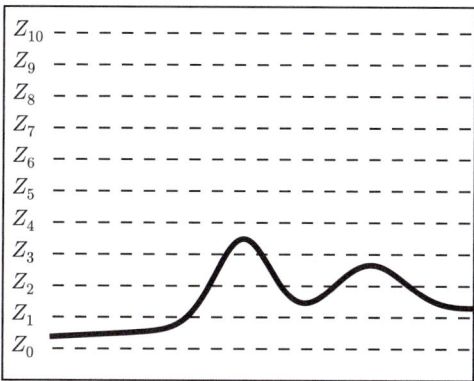

Fig. 2.1: Illustration of height coordinates, showing that coordinate surfaces can intersect the terrain (bold line) and even be underground. This is also true of pressure coordinates.

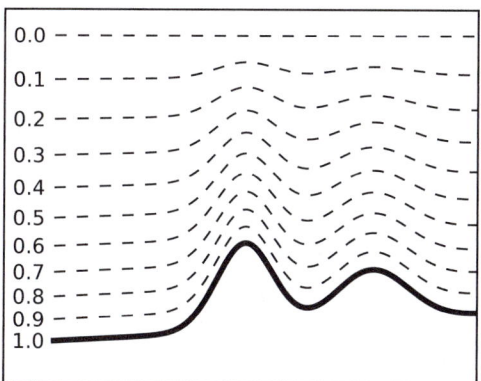

Fig. 2.2: Illustration of sigma coordinates, showing that they never intersect the terrain. The sigma level having the value one is always coincident with the model terrain (bold line), while the sigma level having the value zero is at the top of the model. Sigma surfaces can be greatly tilted in regions of steep terrain, and their vertical separation is not constant.

and horizontal derivatives are converted using

$$\left(\frac{\partial}{\partial x}\right)_z = \left(\frac{\partial}{\partial x}\right)_\sigma - \left(\frac{\partial z}{\partial x}\right)_\sigma \left(\frac{\partial}{\partial z}\right)_x. \qquad (2.52)$$

Sigma coordinates have the advantage of never intersecting the ground. However, this means that they can be greatly tilted in areas of steep terrain. Also, the vertical separation between two sigma coordinate levels varies across the model domain. The advantages of sigma coordinates generally outweigh their disadvantages, and many atmospheric numerical models use some form of sigma coordinates.

2.5   Subgrid-scale Processes: Turbulence, Clouds, Convection, and Radiation

A numerical model solves the governing equations at grid points. If the model is to resolve small-scale phenomena, then a small grid-point spacing is required. If the grid-point spacing is denoted as d, the smallest scale that the model will be able to resolve is $2d$. Processes that occur at length scales smaller than this are known as *subgrid-scale* processes.

One example of a subgrid-scale process is turbulence. Clouds and convection are also subgrid-scale phenomena in all but very high resolution models. They impact the heat balance of the atmosphere through absorption and emission of long and shortwave radiation, as well as through latent heat release. Long and shortwave radiative transfer in clear air are also subgrid-scale processes. Another important process that must be accounted for, especially in longer-range global models, is *gravity wave drag*. This is a process whereby gravity waves are generated near the ground and propagate into the upper atmosphere where they break and act to slow the upper atmosphere flow.

Subgrid-scale processes must be *parameterized* in terms of the model variables. These parameterizations are discussed more fully in Chapter 10.

Exercises

Ex. 2.1: Use the definition of the total derivative from (2.2) and the vector identity, $\nabla \cdot (s\vec{A}) = s\nabla \cdot \vec{A} + \nabla s \cdot \vec{A}$, to show the equivalence of Eqs. (2.3) and (2.4).

Ex. 2.2: Use the definition of potential temperature

$$\theta = T \left(\frac{P_R}{p} \right)^{R_d/c_p},$$ \hfill (2.53)

the ideal gas law, and the definition of the Exner function from (2.25) to show the following equivalencies:

$$-\frac{1}{\rho}\frac{\partial p}{\partial x} = -c_p \theta \frac{\partial \pi}{\partial x}$$

$$-\frac{1}{\rho}\frac{\partial p}{\partial y} = -c_p \theta \frac{\partial \pi}{\partial y}$$

$$-\frac{1}{\rho}\frac{\partial p}{\partial z} = -c_p \theta \frac{\partial \pi}{\partial z}$$

Ex. 2.3: Show that the horizontal pressure gradient acceleration in pressure coordinates is

$$-\frac{1}{\rho}\left(\frac{\partial p}{\partial x}\right)_z = -\left(\frac{\partial \Phi}{\partial x}\right)_p,$$

where Φ is the geopotential ($d\Phi = gdz$). The term $(\partial \Phi / \partial x)_p$ represents the geopotential gradient on a constant pressure surface, and increases in magnitude as the pressure surface becomes more tilted.

Ex. 2.4: The figure below shows pressure surfaces as dotted lines, labeled in hPa. The solid line shows the terrain. The pressure at the top of the model domain is 200 hPa. Calculate the value of σ at each bold dot in the diagram.

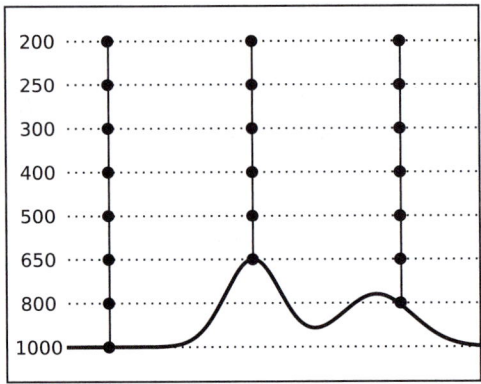

Ex. 2.5: Show that the horizontal pressure gradient acceleration in sigma coordinates is

$$-\frac{1}{\rho}\left(\frac{\partial p}{\partial x}\right)_z = -\frac{1}{\rho}\left(\frac{\partial \tilde{p}}{\partial x}\right)_\sigma + \sigma\frac{\partial \Phi_S}{\partial x} - \left(\frac{\partial \Phi}{\partial x}\right)_\sigma,$$

where Φ_S is the geopotential at the surface of the terrain, and the term $(\partial \Phi/\partial x)_\sigma$ is the geopotential gradient along a constant sigma surface. This last term increases in magnitude as the sigma surface becomes more tilted.

## CHAPTER 3

Finite Differencing

3.1 Difference Equations

The equations of motion contain partial differential equations. To solve them numerically they must be converted into *difference equations* by replacing the derivatives with finite differences. The rationale for replacing a derivative with a finite difference comes directly from the definition of the derivative,

$$\frac{dF}{dt} \equiv \lim_{\Delta t \to 0} \frac{F(t + \Delta t) - F(t)}{\Delta t}. \tag{3.1}$$

The derivative is approximated by ignoring the limit and writing

$$\frac{dF}{dt} \cong \frac{F(t + \Delta t) - F(t)}{\Delta t} = \frac{\Delta F}{\Delta t}. \tag{3.2}$$

Replacing the derivatives by finite differences alters an equation, and the solution to the difference equation will not be exactly equal to the solution of the original differential equation. The hope is that the solution will be "close enough" to provide meaningful information. We expect the errors to be less if a smaller increment for the independent variable (in this case, Δt) is used.

As an illustration, consider the differential equation for the displacement, $F(t)$, of an object undergoing a uniform acceleration, a.

If the initial velocity of the object is zero, then the differential equation for the displacement of the object at any time t is

$$\frac{dF}{dt} = at. \tag{3.3}$$

The analytical solution of (3.3) for an initial displacement of zero is

$$F(t) = \frac{1}{2}at^2. \tag{3.4}$$

The displacement at any time t can be determined from (3.4).

Though finite differences are not needed to solve (3.3), we will use them anyway to illustrate how they are used, and to compare the solution by finite differences with the analytic solution. One finite-difference form of (3.3) is

$$\frac{F(t + \Delta t) - F(t)}{\Delta t} = at, \tag{3.5}$$

which rearranges to

$$F(t + \Delta t) = F(t) + at\,\Delta t. \tag{3.6}$$

A more useful way to write (3.6) is

$$F^{n+1} = F^n + at^n \Delta t, \tag{3.7}$$

where the superscript refers to the value or index of the time step; n is the current time step, while $n + 1$ is the time step being predicted (do not confuse n with an exponent). The displacement at any time t is estimated by starting from the initial time step ($n = 0$) and recursively applying (3.7) for some small time increment Δt. This means that at each time step (3.7) is applied to find the value of F at the next time step. This new value is then used on the right-hand side of (3.7) to find the next new time step, and the process is repeated until the value at the desired, final time step (denoted as N_f) is obtained. This recursive process is illustrated as

$$F^1 = F^0 + at^0 \Delta t$$

$$\searrow$$

$$F^2 = F^1 + at^1 \Delta t$$

$$\searrow$$

$$F^3 = F^2 + at^2 \Delta t$$

$$\vdots$$

$$F^{N_f-1} = F^{N_f-2} + at^{N_f-2} \Delta t$$

$$\searrow$$

$$F^{N_f} = F^{N_f-1} + at^{N_f-1} \Delta t \quad .$$

As an example, we will estimate the displacement after 5 seconds using (3.7), and compare it to the analytical solution from (3.4), using an acceleration of 10 m s^{-2}. We will first use a time increment of 1 sec, thus requiring five iterations to find the value of F after 5 seconds. Table 3.1 compares the analytical solution, $F(t)$, calculated from (3.4) with the finite-difference solution, F^n, from (3.7). When viewing Table 3.1 remember that F^n is the displacement at the current time step n, while F^{n+1} is the displacement at the next time step.

The solution from the difference equation does not agree very well with the analytical solution of the differential equation. We might expect the result to be closer if a smaller time increment is used. Table 3.2 shows the same calculations only for a time increment $\Delta t = 0.5$ seconds (requiring 10 iterations), and indeed a significantly better agreement is achieved.

A further reduction in time increment results in even greater agreement between the analytical and finite-difference solutions, as Table 3.3 illustrates. The results demonstrate that an approximate solution to an ordinary differential equation can be found using finite differences and that, by using a smaller time increment, the accuracy of the approximation can be increased. In the limit, as the time increment becomes infinitesimally small, the approximation can be made to approach the actual solution.[1] However, a smaller time increment comes at the cost of increased number of iterations

---

[1]Since computers cannot represent floating-point numbers to an infinite num-

Table 3.1: Comparison of the finite-difference solution and the analytical solution for $\Delta t = 1$ s. Observe that F^{n+1} at a certain time step is used as F^n at the next time step.

t seconds	n	$F(t)$ meters	F^n meters	F^{n+1} meters	Iteration number
0.0	0	0.0	0.0	0.0	1
1.0	1	5.0	0.0	10.0	2
2.0	2	20.0	10.0	30.0	3
3.0	3	45.0	30.0	60.0	4
4.0	4	80.0	60.0	100.0	5
5.0	5	125.0	100.0		

Table 3.2: Comparison of the finite-difference solution and the analytical solution for $\Delta t = 0.5$ s.

t seconds	n	$F(t)$ meters	F^n meters
0.0	0	0.0	0.0
1.0	2	5.0	2.5
2.0	4	20.0	15.0
3.0	6	45.0	37.5
4.0	8	80.0	70.0
5.0	10	125.0	112.5

Table 3.3: Comparison of the analytical solution and finite-difference solutions for varying time increments.

t seconds	$F(t)$ meters	F^n (meters) for various time increments			
		$\Delta t = 1$ s	$\Delta t = 0.5$ s	$\Delta t = 0.1$ s	$\Delta t = 0.01$ s
0.0	0.0	0.0	0.0	0.0	0.0
1.0	5.0	0.0	2.5	4.7	5.0
2.0	20.0	10.0	15.0	19.2	19.9
3.0	45.0	30.0	37.5	43.7	44.9
4.0	80.0	60.0	70.0	78.2	79.8
5.0	125.0	100.0	112.5	115.7	124.8
(Total iterations)		(5)	(10)	(50)	(500)

required. Calculating the results for a time increment of 0.01 seconds requires 100 times as many computations as for a 1 second time increment.

Using finite differences to solve (3.3) was merely an academic exercise, since an analytical solution is easily found. However, there are numerous differential equations that have no analytical solution and for which the only hope of solution rests with numerical methods such as finite differences.

3.2 Forward Differencing

Section 3.1 gave a general overview of finite differences and how they are used. In this section we discuss the mathematical basis for finite differences and ways to assess the accuracy of the approximations.

Fundamental to the theory of finite differences is the Taylor series, which allows the value of a function, $F(x + \Delta x)$, to be expressed as an infinite series involving the value of the function and its derivatives at x. The formula for a Taylor series is

$$F(x + \Delta x) = F(x) + \frac{1}{1!}\frac{dF}{dx}\Delta x + \frac{1}{2!}\frac{d^2F}{dx^2}\Delta x^2$$
$$+ \frac{1}{3!}\frac{d^3F}{dx^3}\Delta x^3 + \cdots + \frac{1}{m!}\frac{d^mF}{dx^m}\Delta x^m + \cdots . \quad (3.8)$$

Notice that (3.8) is an *exact* representation for $F(x + \Delta x)$, not an approximation. This is true as long as an infinite number of terms are retained on the right-hand side (RHS). Solving (3.8) for the first derivative shows that

$$\frac{dF}{dx} = \frac{F(x + \Delta x) - F(x)}{\Delta x} - \frac{1}{2!}\frac{d^2F}{dx^2}\Delta x$$
$$- \frac{1}{3!}\frac{d^3F}{dx^3}\Delta x^2 - \cdots - \frac{1}{m!}\frac{d^mF}{dx^m}\Delta x^{m-1} - \cdots . \quad (3.9)$$

Again, the RHS of (3.9) is exactly equivalent to the first derivative so long as an infinite number of terms are included (or all the terms for which the function F has a non-zero n^{th} derivative).

---

ber of decimal places, but instead must truncate them, roundoff error becomes an issue for extremely small time increments. Beyond this point, further reduction of time increment may actually result in decreased accuracy.

An approximation to the first derivative can be made by neglecting all but the first term on the RHS,

$$\frac{dF}{dx} \cong \frac{F(x + \Delta x) - F(x)}{\Delta x}. \qquad (3.10)$$

This is the same approximation for the derivative that we used in the example in the previous section and is called a *forward difference* because we used information at both x and $x + \Delta x$. The error in making this approximation is equal to the sum of the terms that were ignored on the RHS of (3.9). These terms are the *error terms*, and are

$$\varepsilon = -\frac{1}{2!}\frac{d^2F}{dx^2}\Delta x - \frac{1}{3!}\frac{d^3F}{dx^3}\Delta x^2 - \cdots - \frac{1}{m!}\frac{d^mF}{dx^m}\Delta x^{m-1} - \cdots. \qquad (3.11)$$

The largest-magnitude term in the error is the first term, and it is first-order with respect to Δx (the exponent on Δx is equal to one). Therefore, the approximation of a first derivative by the forward difference, (3.10), is said to be *first-order* accurate.

3.3   Other Finite-Difference Representations

There are many possible finite-difference approximations for the first derivative. A forward difference was illustrated in (3.10). Backward differencing, centered differencing, and fourth-order differencing are some other representations for a first derivative.

3.3.1   Backward differencing

Equation (3.10) is called a forward difference because the derivative at x is approximated using information at a higher value of x (the point *ahead* or *forward* of x). A Taylor series for a negative (or *backward*) interval can be written as

$$F(x - \Delta x) = F(x) - \frac{1}{1!}\frac{dF}{dx}\Delta x + \frac{1}{2!}\frac{d^2F}{dx^2}\Delta x^2 - \frac{1}{3!}\frac{d^3F}{dx^3}\Delta x^3 + \cdots$$
$$+ (-1)^m \frac{1}{m!}\frac{d^mF}{dx^m}\Delta x^m + \cdots \qquad (3.12)$$

and the first derivative written as

$$\frac{dF}{dx} = \frac{F(x) - F(x - \Delta x)}{\Delta x} + \frac{1}{2!}\frac{d^2F}{dx^2}\Delta x - \frac{1}{3!}\frac{d^3F}{dx^3}\Delta x^2 + \cdots$$
$$+ (-1)^m \frac{1}{m!}\frac{d^mF}{dx^m}\Delta x^{m-1} + \cdots. \quad (3.13)$$

In this case, the first derivative can be approximated as

$$\frac{dF}{dx} \cong \frac{F(x) - F(x - \Delta x)}{\Delta x}, \quad (3.14)$$

which is also first-order accurate. Equation (3.14) is known as a *backward difference* because the derivative at x is approximated using information at a lower value of x (the point *behind x*).

3.3.2 Centered differencing

Another possibility is to subtract (3.12) from (3.8) and solve for the first derivative, yielding

$$\frac{dF}{dx} = \frac{F(x + \Delta x) - F(x - \Delta x)}{2\Delta x} - \frac{d^3F}{dx^3}\frac{\Delta x^2}{3!} - \frac{d^5F}{dx^5}\frac{\Delta x^4}{5!} - \cdots. \quad (3.15)$$

In this case, we have a *centered* difference,

$$\frac{dF}{dx} \cong \frac{F(x + \Delta x) - F(x - \Delta x)}{2\Delta x}. \quad (3.16)$$

The centered difference is *second-order* accurate, because the first-order terms in the error canceled so that the largest term in the error involves Δx^2. This increase in accuracy makes the centered difference a widely used choice for approximating spatial first derivatives in atmospheric numerical models.

3.3.3 Fourth-order differencing

Another second-order accurate representation for a first derivative is

$$\frac{dF}{dx} = \frac{F(x + 2\Delta x) - F(x - 2\Delta x)}{4\Delta x}$$
$$- \frac{d^3F}{dx^3}\frac{(2\Delta x)^2}{3!} - \frac{d^5F}{dx^5}\frac{(2\Delta x)^4}{5!} - \cdots. \quad (3.17)$$

This is nothing more than the centered difference using a space interval of $2\Delta x$ rather than Δx. If we could combine (3.15) and (3.17) such that the second-order terms (those containing Δx^2) vanish, then we would have a fourth-order accurate scheme. We do this by multiplying (3.15) by a, and (3.17) by b, to get

$$
(a + b)\frac{dF}{dx} = a\frac{F(x + \Delta x) - F(x - \Delta x)}{2\Delta x}
$$
$$
+ b\frac{F(x + 2\Delta x) - F(x - 2\Delta x)}{4\Delta x}
$$
$$
- (a + 4b)\frac{d^3 F}{dx^3}\frac{\Delta x^2}{3!} - (a + 16b)\frac{d^5 F}{dx^5}\frac{\Delta x^4}{5!} - \cdots . \quad (3.18)
$$

If values for a and b are chosen such that

$$
a + b = 1
$$
$$
a + 4b = 0 \quad\quad (3.19)
$$

then the second-order term would disappear, and a fourth-order accurate scheme,

$$
\frac{dF}{dx} = a\frac{F(x + \Delta x) - F(x - \Delta x)}{2\Delta x} + b\frac{F(x + 2\Delta x) - F(x - 2\Delta x)}{4\Delta x}
$$
$$
- (a + 16b)\frac{d^5 F}{dx^5}\frac{\Delta x^4}{5!} - \cdots , \quad (3.20)
$$

would result. The values for a and b are found by solving (3.19), resulting in $a = 4/3$, and $b = -1/3$. The fourth-order scheme is then

$$
\frac{dF}{dx} \cong \frac{4}{3}\left[\frac{F(x + \Delta x) - F(x - \Delta x)}{2\Delta x}\right]
$$
$$
- \frac{1}{3}\left[\frac{F(x + 2\Delta x) - F(x - 2\Delta x)}{4\Delta x}\right] . \quad (3.21)
$$

3.4   Second Derivatives

Finite-difference approximations to second derivatives are also constructed from the Taylor series expansions for $F(x + \Delta x)$ and $F(x - \Delta x)$. These expansions, repeated here, are

$$F(x + \Delta x) = F(x) + \frac{1}{1!}\frac{dF}{dx}\Delta x + \frac{1}{2!}\frac{d^2F}{dx^2}\Delta x^2 + \cdots$$
$$+ \frac{1}{m!}\frac{d^mF}{dx^m}\Delta x^m + \cdots \quad (3.22)$$

$$F(x - \Delta x) = F(x) - \frac{1}{1!}\frac{dF}{dx}\Delta x + \frac{1}{2!}\frac{d^2F}{dx^2}\Delta x^2 + \cdots$$
$$+ (-1)^m \frac{1}{m!}\frac{d^mF}{dx^m}\Delta x^m + \cdots. \quad (3.23)$$

Adding these two expansions together, we get

$$F(x + \Delta x) + F(x - \Delta x) = 2F(x) + \frac{d^2F}{dx^2}\Delta x^2 + \frac{2}{4!}\frac{d^4F}{dx^4}\Delta x^4$$
$$+ \frac{2}{6!}\frac{d^6F}{dx^6}\Delta x^6 + \cdots + \frac{2}{m!}\frac{d^mF}{dx^m}\Delta x^m + \cdots \quad (3.24)$$

where m is even. This can be rearranged to give

$$\frac{d^2F}{dx^2} \cong \frac{F(x + \Delta x) + F(x - \Delta x) - 2F(x)}{\Delta x^2}, \quad (3.25)$$

a second-order accurate approximation for the second derivative. A fourth-order approximation for the second derivative can also be derived in a manner similar to that of the fourth-order scheme for the first derivative. This derivation is not shown, but the result is

$$\frac{d^2F}{dx^2} \cong \frac{1}{\Delta x^2}\left\{ \frac{4}{3}\left[F(x + \Delta x) + F(x - \Delta x)\right]\right.$$
$$\left. - \frac{1}{12}\left[F(x + 2\Delta x) + F(x - 2\Delta x)\right] - \frac{5}{2}F(x) \right\}. \quad (3.26)$$

3.5   Time-differencing Schemes

Finite differences can be used to approximate derivatives either in time or space. Though the mathematical basis is the same in either case, the application of finite differences to time derivatives deserves special consideration. In this section we discuss three of the most fundamental time-differencing schemes applied to a general, first-order differential equation. In Chapter 4 we will apply these schemes to specific differential equations that are often encountered in atmospheric modeling.

3.5.1   Forward (Euler) scheme

Imagine a general, first-order differential equation for some function, $F(t)$,

$$\frac{dF}{dt} = G(F), \tag{3.27}$$

where G is some completely general function of F (and hence, also a function of t). Some examples include:

$$\frac{dF}{dt} = F(t); \qquad G = F(t)$$

$$\frac{dF}{dt} = [F(t)]^2; \qquad G = [F(t)]^2 \tag{3.28}$$

$$\frac{dF}{dt} = \sin[F(t)]; \qquad G = \sin[F(t)].$$

The finite-difference form of (3.27), using a forward difference for the derivative, is

$$\frac{F^{n+1} - F^n}{\Delta t} = G^n, \tag{3.29}$$

which rearranges to

$$F^{n+1} = F^n + G^n \Delta t. \tag{3.30}$$

Equation (3.30) illustrates why this is called a *forward-in-time* scheme, since we are using information at the current time step (F^n and G^n) to find the value of F at the next time step (F^{n+1}). This scheme is also called the *Euler* scheme, after the mathematician Leonhard Euler.

The Euler scheme is the most straightforward and intuitive time differencing scheme, which is why we used it in the example in Section 3.1. Unfortunately, it has serious drawbacks when applied to the equations of motion, as we will find in Chapter 4. Therefore, we explore other possible time differencing schemes.

3.5.2 Backward scheme

If the time derivative in (3.27) is approximated using a backward differencing scheme

$$\frac{dF}{dt} \cong \frac{F^n - F^{n-1}}{\Delta t}, \tag{3.31}$$

the recursive equation becomes

$$F^n = F^{n-1} + G^n \Delta t,$$

which is equivalent to

$$F^{n+1} = F^n + G^{n+1} \Delta t. \tag{3.32}$$

Though this looks similar to (3.30), it is substantially different. In order to calculate the new value of F at the next time step, F^{n+1}, *we are required to already know* F^{n+1}, since G^{n+1} is a function of F^{n+1}! A scheme such as this, where the value of the dependent variable at the future time step is needed on the right-hand side of the recursive relation, is known as an *implicit* scheme (in contrast to the forward scheme, which is *explicit*.)

At first, it might seem that an implicit scheme is merely an academic oddity; of what use is a scheme that requires us to already know the answer in order to find the answer? In Chapter 4 we will find that implicit schemes, such as (3.32), are indeed solvable, though not in a straight-forward manner. They are frequently used in numerical modeling, and actually have certain advantages over explicit schemes.

3.5.3 Centered (leapfrog) scheme

Another possibility for representing the time derivative in (3.27) is to use *centered differencing*, approximating the time derivative as

$$\frac{dF}{dt} \cong \frac{F^{n+1} - F^{n-1}}{2\Delta t}, \tag{3.33}$$

yielding the recursive equation

$$F^{n+1} = F^{n-1} + 2G^n \Delta t. \tag{3.34}$$

This scheme is widely known as the *leapfrog scheme*, since the solution leaps from the previous time step to the future time step, by-passing the current time step. The leapfrog scheme is also explicit, since information from time step $n + 1$ is not found on the RHS of the equation. It is also an example of a *two-level scheme*, since information from two time steps (n and $n - 1$) is needed to predict the value at the future time step. The leapfrog scheme is more accurate than either the forward or backward schemes, and is therefore widely used in atmospheric numerical models. It is not without disadvantages, however, as we shall find in Chapter 4.

Exercises

Note: Exercises 3.1 and 3.2 are well-suited to solution using either a computer program or calculations within a spreadsheet program. However, they may be solved by hand as well.

Ex. 3.1: Start with the cubic polynomial $F(x) = ax^3 + bx^2 + cx + d$. Pick any non-zero values you like for the coefficients a, b, c, and d.
a. Calculate the first derivative at several arbitrary values of x using forward, backward, and centered finite differencing and a value of Δx of your choosing. Compare your calculated values to the analytical first derivative. Which scheme comes closest to the analytical values?
b. For one or two of the x values you chose in part **a**, recalculate the first derivative using centered finite differencing for smaller and smaller values of Δx. How does the difference between the analytical and calculated values change as Δx becomes smaller?

Ex. 3.2: Start with the quadratic polynomial $F(x) = ax^2 + bx + c$. Pick any non-zero values you like for the coefficients a, b, and c. Calculate the first derivative at several arbitrary values of x using the centered finite-difference scheme and several different values of Δx. Compare your calculated values to the analytical first derivative. Why are your calculated values so close to the analytical value regardless of what value you choose for Δx? Hint: Study (3.15) carefully.

Ex. 3.3: For the function $F(x) = 400 \cos(\pi x / 16)$

a. Calculate the first derivative at $x = 3$ using forward, backward, centered, and fourth-order differencing with $\Delta x = 1$. Compare the results to the analytical derivative and see which scheme is closest. *Note*: Carry your calculations to at least three decimal places to ensure the differences are apparent.

b. Calculate the second derivative at $x = 3$ using the second-order and fourth-order differencing with $\Delta x = 1$. Compare the results to the analytical derivative and see which scheme is closest. *Note*: Carry your calculations to at least four decimal places to ensure the differences are apparent.

Ex. 3.4: Show that centered differencing for the first derivative is essentially the average of the forward and backward difference.

Ex. 3.5: Derive the fourth-order scheme for the second derivative, (3.26).

## CHAPTER 4

Application of Finite Differencing

4.1 Grid Points and Notation

In Chapter 3 we discussed finite-difference methods that can be used to approximate first and second derivatives. The three main techniques for first derivatives were the forward scheme, the backward scheme, and the centered scheme (also known as the leapfrog scheme if used to approximate a time derivative). We now want to apply these schemes to a time-dependent partial differential equation (PDE) in order to approximate the future solutions to the equation. Before we begin we need to establish the notation that will be used regarding grid points and time steps.

In order to solve a differential equation using finite differences we need to know the values of the dependent variable u at discrete points in space. These discrete points are known as *grid points*. For simplicity we will assume the grid points are evenly spaced, and that the grid increment or grid spacing (the distance between adjacent grid points) has a value of Δx. Each grid point is uniquely numbered with an integer value, i. Figure 4.1 illustrates a one-dimensional grid.

The solution to the partial differential equation is denoted as $u(x, t)$. The solution to the corresponding finite-difference equation

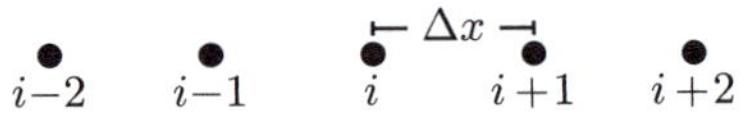

Fig. 4.1: A one-dimensional grid of grid spacing Δx.

will be written as u_i^n where the subscript i refers to the index of the grid point, and the superscript n refers to the index of the time step. The time step index begins at zero, which corresponds to the initial condition for the equation. The beginning value for the grid point index is also zero.[1] Recall that u_i^n is an approximation for $u(x, t)$. Table 4.1 illustrates the correspondence between the two notations.

If the partial differential equation were two- or three-dimensional, there would be additional subscripts used, one for each dimension. In this case the correspondence between the two notations would be $u(x, y, t) \rightarrow u_{i,j}^n$ or $u(x, y, z, t) \rightarrow u_{i,j,k}^n$.

4.2 Grid Spacing and Grid Resolution

A discrete grid can resolve only those features that are above a certain size. Since any feature can be decomposed through Fourier analysis into sinusoids of varying amplitude, wavelength, and phase, there is a smallest resolvable wavelength given by

$$\lambda_{\min} = 2\Delta x. \tag{4.1}$$

Alternately, we can say that the maximum resolvable wave number is

$$k_{\max} = 2\pi / \lambda_{\min} = \pi / \Delta x. \tag{4.2}$$

Features whose wavelength is smaller than $\lambda_{\min}$ (wave number greater than $k_{\max}$) will be aliased to longer wavelengths (smaller wave numbers). The grid spacing, Δx, is sometimes inappropriately referred to as the grid resolution. However, the resolution is

---

[1]The choice of zero for the beginning time step and grid point index is rather arbitrary. Either or both may also be set as one. The ambiguity stems from the fact that in the Fortran programming language, array indices begin with one, while in other common programming languages (such as C++ and Python), array indices begin with zero.

Table 4.1: Correspondence of notation used for solution to the partial differential equation (PDE) and the finite-difference equation (FDE) for a space increment of $\Delta x = 2$ meters and a time increment of $\Delta t = 30$ seconds.

Solution to PDE	Solution to FDE
$u(x, t)$	u_i^n
$u(6 \text{ m}, t)$	u_3^n
$u(x, 900 \text{ s})$	u_i^{30}
$u(0 \text{ m}, 0 \text{ s})$	u_0^0
$u(4 \text{ m}, 180 \text{ s})$	u_2^6
$u(0 \text{ m}, 180 \text{ s})$	u_0^6
$u(4 \text{ m}, 0 \text{ s})$	u_2^0

more appropriately described as the wavelength, $\lambda_{\min}$, of the smallest wave that can represented by the grid. The term *grid resolution* is therefore twice the grid spacing.

4.3 The Advection Equation

The choice of which finite-differencing scheme to apply to an equation depends on the equation. Some schemes cannot be used for approximating time derivatives in certain equations because the solution grows exponentially with time. Such schemes are said to be unstable for that particular equation. However, a scheme that is unstable when applied to one equation may be stable when used on a different equation.

The simplest way to illustrate the time-differencing and space-differencing schemes we have discussed so far is to apply them to the linearized *advection equation* (also called the *one-way wave equation*) given by

$$\frac{\partial u}{\partial t} + c\frac{\partial u}{\partial x} = 0. \tag{4.3}$$

The variable u is a general scalar, and is not necessarily the x-component of velocity (although it could be). A form of this equation appears in the equations of motion in the advection terms. The

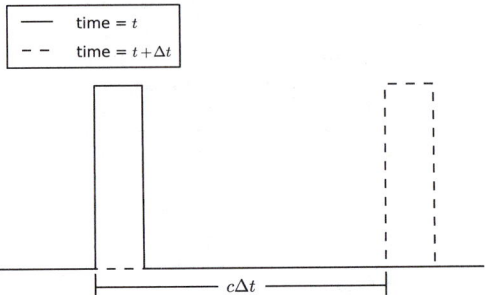

Fig. 4.2: A box-shaped solution to (4.3) for a positive value of c. The solution translates to the right while preserving its shape and amplitude.

general solution to (4.3) is $u = f(x - ct)$, where f is any arbitrary function. The solutions preserve their shape and amplitude; a triangle will translate as a triangle, a square will translate as a square, and any arbitrary shape, regardless of how complex, will translate at speed c without changing form. A possible solution to (4.3) is illustrated in Fig. 4.2.

Since almost any arbitrary function can be decomposed through Fourier analysis into a summation of pure sinusoids, it is helpful to use a pure sinusoidal wave when analyzing (4.3). A pure sinusoidal wave is written as

$$u(x,t) = A(t)e^{\iota kx}, \qquad (4.4)$$

where $A(t)$ is the complex amplitude, and k is the wave number. We use the bold Greek letter iota (ι) to represent the imaginary number $\sqrt{-1}$. The complex amplitude is given by

$$A(t) = ae^{-\iota \omega t} \qquad (4.5)$$

where ω is the angular frequency of the sinusoidal component and a is constant. A key point is that although $A(t)$ changes with time, its magnitude remains constant.[2]

---

[2] $|A(t)| = |a\exp(-\iota \omega t)| = |a\cos \omega t - \iota a \sin \omega t| = \sqrt{a^2(\cos^2 \omega t + \sin^2 \omega t)} = a$

Table 4.2: Some possibilities for the finite-difference form of the advection equation.

	Forward-in-time	*Centered-in-time*	*Backward-in-time*
Forward-in-space	$\dfrac{u_i^{n+1}-u_i^n}{\Delta t} + c\dfrac{u_{i+1}^n-u_i^n}{\Delta x} = 0$	$\dfrac{u_i^{n+1}-u_i^{n-1}}{2\Delta t} + c\dfrac{u_{i+1}^n-u_i^n}{\Delta x} = 0$	$\dfrac{u_i^{n+1}-u_i^n}{\Delta t} + c\dfrac{u_{i+1}^{n+1}-u_i^{n+1}}{\Delta x} = 0$
Centered-in-space	$\dfrac{u_i^{n+1}-u_i^n}{\Delta t} + c\dfrac{u_{i+1}^n-u_{i-1}^n}{2\Delta x} = 0$	$\dfrac{u_i^{n+1}-u_i^{n-1}}{2\Delta t} + c\dfrac{u_{i+1}^n-u_{i-1}^n}{2\Delta x} = 0$	$\dfrac{u_i^{n+1}-u_i^n}{\Delta t} + c\dfrac{u_{i+1}^{n+1}-u_{i-1}^{n+1}}{2\Delta x} = 0$
Backward-in-space	$\dfrac{u_i^{n+1}-u_i^n}{\Delta t} + c\dfrac{u_i^n-u_{i-1}^n}{\Delta x} = 0$	$\dfrac{u_i^{n+1}-u_i^{n-1}}{2\Delta t} + c\dfrac{u_i^n-u_{i-1}^n}{\Delta x} = 0$	$\dfrac{u_i^{n+1}-u_i^n}{\Delta t} + c\dfrac{u_i^{n+1}-u_{i-1}^{n+1}}{\Delta x} = 0$

4.4   Finite Differencing Applied to the Advection Equation

When applying finite differencing to the advection equation we have a choice of using forward, centered, or backward differencing for both the time and space derivatives. Table 4.2 shows there are at least nine different possibilities for writing the advection equation in finite-difference form. In the following sections we will rigorously analyze each of these choices and see that they are not all equally desirable. Some of them are unstable and have solutions that quickly grow exceedingly large.

4.4.1   Forward-in-time, backward-in-space differencing

The forward-in-time, backward-in-space scheme applied to the advection equation is

$$\frac{u_i^{n+1} - u_i^n}{\Delta t} + c\frac{u_i^n - u_{i-1}^n}{\Delta x} = 0. \qquad (4.6)$$

For positive values of c this scheme is also known as the *upwind scheme*, since the space derivative at a given grid point is approxi-

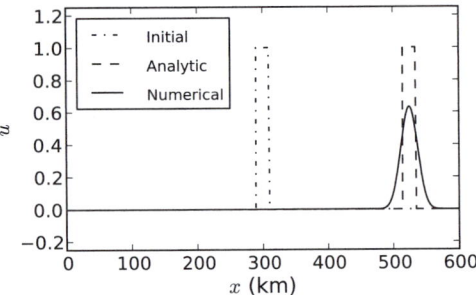

Fig. 4.3: The initial value of u, and the analytic and numerical solutions for the advection equation after 500 time steps. Numerical solution used the *forward-in-time, backward-in-space* scheme with values of: $c = 15$ m s^{-1}; $\Delta x = 1$ km; and $\Delta t = 30$ s.

mated by using information from that grid point and the adjacent *upwind* grid point. For negative values of c this scheme is known as the *downwind scheme*. Equation (4.6) can be written in recursive form as

$$u_i^{n+1} = u_i^n - \frac{c\Delta t}{\Delta x}\left(u_i^n - u_{i-1}^n\right). \tag{4.7}$$

At each time step, n, all the information on the right-hand side of (4.7) is known, and so the solution at time level $n+1$ can be found.

Figure 4.3 shows the initial values of u at each grid point and the results at time step $n = 500$ using (4.7). The exact solution to the advection equation preserves its shape and amplitude. The numerical solution from the upwind scheme neither preserves shape nor amplitude. The solution after 500 time steps is much more rounded and also of lower amplitude. The scheme therefore does not replicate the exact solution to the differential equation. This is a general property of all finite-difference schemes - *they will never replicate the exact solution to a differential equation!* The hope is that the solution to the finite-difference equation will at least be close enough to the exact solution of the differential equation that it will give meaningful information.

Figure 4.4 shows the result of using the forward-in-time, backward-in-space scheme for the same initial data as in Fig. 4.3, but with a time increment that is three times longer (90 s). In this case, the solution is seen to rapidly increase with time, achieving an

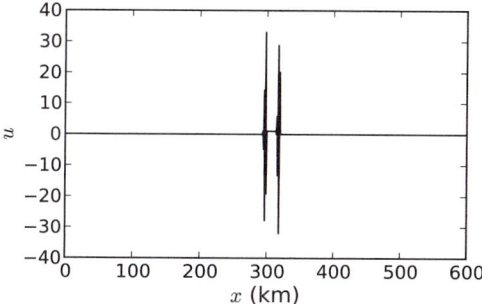

Fig. 4.4: Numerical solution to the advection equation after only 10 time steps using the *forward-in-time, backward-in-space* finite-difference form of the advection equation, using a time interval three times longer (90 s) than in Fig. 4.3. The numerical solution looks nothing like the analytic solution.

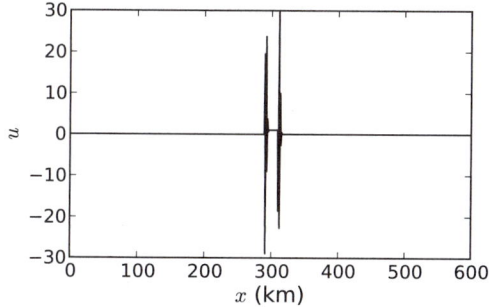

Fig. 4.5: Numerical solution to the advection equation after only 8 time steps, using the downwind scheme. All parameters are the same as for Fig. 4.3, except that the propagation speed, c, is -15 m s^{-1}.

amplitude over 30 times greater than the initial value in just 10 time steps. This behavior is known as 'blowing up', and means that the scheme is *unstable*. Fig. 4.5 shows the result of using the forward-in-time, backward-in-space scheme with a negative value of c (this is known as a *downwind* scheme). The downwind scheme is also seen to be unstable.

4.4.2   Analyzing stability

The results of the prior section show that the forward-in-time, backward-in-space scheme may be stable or unstable, depending on the parameters chosen for c and Δt. There is a mathematical way to analyze the stability of a scheme. This section illustrates the method for the forward-in-time, backward-in-space scheme. We will then apply it to a number of other schemes.

The first step in analyzing the stability of a finite-difference scheme is to write a general solution to the finite-difference equation, in this case (4.6) or (4.7). Since this equation is a discrete form of the advection equation, the solution will be a discrete form of (4.5). This solution has the form[3]

$$u_i^n = \lambda^n A e^{\imath k i \Delta x}. \tag{4.8}$$

In this representation A is a complex constant and λ is a complex number called the *amplification factor*. The amplification factor describes how the amplitude of the solution changes from one time level to the next. If the solution is neither growing nor decaying with time, then λ will be equal to one. The superscript n on λ behaves like an exponent, so that $\lambda^{n+1} = \lambda^n \lambda$.

Since (4.6) is only an approximation to (4.3), its solution given by (4.8) is expected to only be an approximation to the actual solution of the advection equation. It is important to know how this approximate solution behaves, and to do this we need to know the value of the amplification factor. We find this by substituting the discrete solution from (4.8) into either (4.6) or (4.7) to get

$$\lambda^{n+1} A e^{\imath k i \Delta x} = \lambda^n A e^{\imath k i \Delta x} - \frac{c \Delta t}{\Delta x} \left(\lambda^n A e^{\imath k i \Delta x} - \lambda^n A e^{\imath k (i-1) \Delta x} \right), \tag{4.9}$$

which upon factoring $A e^{\imath k i \Delta x}$ from each term simplifies to

$$\lambda^{n+1} = \lambda^n - \frac{c \Delta t}{\Delta x} \left[1 - e^{-\imath k \Delta x} \right] \lambda^n. \tag{4.10}$$

Dividing through by λ^n, we get

$$\lambda = 1 - \frac{c \Delta t}{\Delta x} \left[1 - e^{-\imath k \Delta x} \right]. \tag{4.11}$$

Using Euler's formula,

$$e^{-\imath k \Delta x} = \cos k \Delta x - \imath \sin k \Delta x, \tag{4.12}$$

---

[3]Care must be exercised to avoid confusing the imaginary number $\imath$ with the grid-point index i.

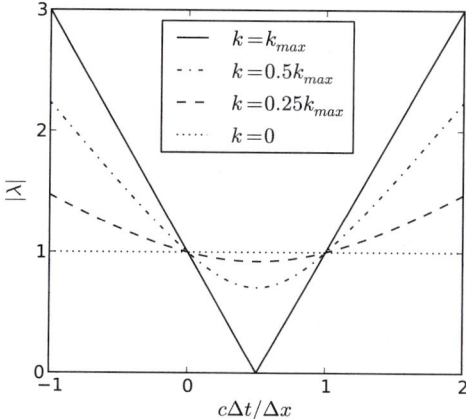

Fig. 4.6: Magnitude of amplification factor for the forward-in-time, backward-in-space scheme for $k = 0$, $0.25k_{max}$, $0.5k_{max}$, and k_{max}. Scheme is unstable for all negative values of c and for any values of c such that $c\Delta t/\Delta x > 1$.

(4.11) becomes

$$\lambda = 1 - \frac{c\Delta t}{\Delta x}\left(1 - \cos k\Delta x\right) + \imath\frac{c\Delta t}{\Delta x}\sin k\Delta x. \qquad (4.13)$$

The amplification factor is a complex number. The magnitude of the amplification factor indicates whether the solution to the finite-difference equation is growing, decaying, or remaining constant with time. If $|\lambda|$ is greater than one, the solution is growing larger at each time step and will eventually grow to infinity. If this is the case, the solution to the difference equation is *unstable*. If $|\lambda|$ is less than one the solution is decaying with time. Ideally, we would like $|\lambda|$ to be equal to one, so that the amplitude of the solution remains constant with time.

Figure 4.6 shows a plot of the magnitude of the amplification factor as a function of the quantity $c\Delta t/\Delta x$ for four different wave numbers ($k = 0$, $k = 0.25k_{max}$, $k = 0.5k_{max}$, and $k = k_{max}$). This figure shows that the scheme is stable only if the following condition is met:

$$0 \le \frac{c\Delta t}{\Delta x} \le 1. \qquad (4.14)$$

We can now explain the behavior exhibited in Figs. 4.4 and 4.5. When a time increment of 90 seconds was used, the value of $c\Delta t/\Delta x$

exceeded one, and the amplification factor's magnitude was larger than one, resulting in an unstable solution. Likewise, when a negative value of c was used, $c\Delta t/\Delta x$ was negative, again resulting in an unstable solution. We can also explain why the numerical solution in Fig. 4.3 is more rounded and of smaller amplitude than the analytical solution. Figure 4.6 shows that within the stable range the higher wave numbers (shorter wavelengths) are more highly damped than are the lower wave numbers. Since the higher wave numbers are needed to achieve the sharp corners of the original solution, and they are damped with time, the numerical solution becomes more rounded and of less amplitude.

The method we used to analyze the stability of a finite-difference scheme can be summarized as follows:

1. Write the differential equation in finite-difference form.

2. Substitute a solution in the form of (4.8) into the finite-difference equation.

3. Rearrange the resulting equation to solve for the amplification factor, λ.

4. Analyze the conditions under which the magnitude of the amplification factor, $|\lambda|$, exceeds a value of one (for instability) or is less than one (for stability).

4.4.3 Forward-in-time, forward-in-space scheme

The finite-difference form of the advection equation using the forward-in-time, forward-in-space differencing scheme is

$$\frac{u_i^{n+1} - u_i^n}{\Delta t} + c\frac{u_{i+1}^n - u_i^n}{\Delta x} = 0. \qquad (4.15)$$

Note that this scheme differs from the forward-in-time, backward-in-space scheme, (4.6), in that for positive values of c, (4.15) is a downwind scheme while for negative values of c, it is an upwind scheme. Equation (4.15) yields the recursive equation

$$u_i^{n+1} = u_i^n - \frac{c\Delta t}{\Delta x}\left(u_{i+1}^n - u_i^n\right). \qquad (4.16)$$

The stability of this scheme is analyzed by substituting (4.8) into (4.16) and solving for the magnitude of the amplification factor. The resulting stability condition for this scheme (see Exercise 4.1) is

$$-1 \leq \frac{c\Delta t}{\Delta x} \leq 0. \tag{4.17}$$

This scheme is unstable for all positive values of c, and is stable for negative values of c as long as they fall into the range of values such that condition (4.17) is met.

4.4.4   Upwind versus downwind scheme

There is symmetry in the stability conditions for the forward-in-time, backward-in-space scheme and the forward-in-time, forward-in-space scheme. Comparing (4.14) and (4.17) we see that the forward-in-time, backward-in-space scheme is unstable for all negative values of c, whereas the forward-in-time, forward-in-space scheme is unstable for all positive values of c. Thus, either scheme is always unstable when employed as a downwind scheme (i.e., using values at the grid point of interest and at the grid point downwind from the grid point of interest when computing the spatial finite difference). When employed as an upwind scheme (i.e., using values at the grid point of interest and at the grid point upwind from the grid point of interest) either scheme is stable as long as the following condition for stability is met,

$$\left| \frac{c\Delta t}{\Delta x} \right| \leq 1. \tag{4.18}$$

These results are summarized in Table 4.3.

The stability condition (4.18) appears in the stability analysis of other schemes besides the two that we have analyzed so far. It is a fundamental criterion for the stability of many schemes, and is known as the *Courant-Friedrichs-Lewy* stability condition, often referred to simply as the *CFL condition*.

4.4.5   Forward-in-time, centered-in-space scheme

Another possibility for approximating the advection equation with finite differences is to use the forward-in-time, centered-in-space scheme,

$$\frac{u_i^{n+1} - u_i^n}{\Delta t} + c\frac{u_{i+1}^n - u_{i-1}^n}{2\Delta x} = 0, \tag{4.19}$$

Table 4.3: Stability criteria for forward-in-time, forward-in-space and for forward-in-time, backward-in-space schemes.

Forward-in-time, Backward-in-space	$\dfrac{u_i^{n+1}-u_i^n}{\Delta t} +$ $c\dfrac{u_i^n-u_{i-1}^n}{\Delta x} = 0$	$c > 0$ upwind	stable if $\lvert c\Delta t/\Delta x\rvert \leq 1$
		$c < 0$ downwind	always unstable
Forward-in-time, Forward-in-space	$\dfrac{u_i^{n+1}-u_i^n}{\Delta t} +$ $c\dfrac{u_{i+1}^n-u_i^n}{\Delta x} = 0$	$c < 0$ upwind	stable if $\lvert c\Delta t/\Delta x\rvert \leq 1$
		$c > 0$ downwind	always unstable

which can also be written as the recursive relation

$$u_i^{n+1} = u_i^n - \frac{c\Delta t}{2\Delta x}\left(u_{i+1}^n - u_{i-1}^n\right). \tag{4.20}$$

To find how this scheme behaves we substitute (4.8) into (4.20) to get

$$\lambda^{n+1}Ae^{\iota ki\Delta x} = \lambda^n Ae^{\iota ki\Delta x} - \frac{c\Delta t}{2\Delta x}\left(\lambda^n Ae^{\iota k(i+1)\Delta x} - \lambda^n Ae^{\iota k(i-1)\Delta x}\right) \tag{4.21}$$

which simplifies to

$$\lambda = \frac{\lambda^{n+1}}{\lambda^n} = 1 - \frac{c\Delta t}{2\Delta x}\left(e^{\iota k\Delta x} - e^{-\iota k\Delta x}\right). \tag{4.22}$$

From Euler's formula we have

$$e^{\iota k\Delta x} - e^{-\iota k\Delta x} = 2\iota \sin k\Delta x \tag{4.23}$$

so that (4.22) becomes

$$\lambda = 1 - \iota\frac{c\Delta t}{\Delta x}\sin k\Delta x. \tag{4.24}$$

The magnitude of the amplification factor is

$$|\lambda| = \sqrt{1 + \left(\frac{c\Delta t}{\Delta x}\right)^2 \sin^2 k\Delta x}. \tag{4.25}$$

Equation (4.25) is always greater than unity, so the forward-in-time, centered-in-space scheme applied to the advection equation is *always unstable* and should never be used for the advection equation.

4.4.6 The centered-in-time (leapfrog), centered-in-space scheme

We now analyze the stability of the centered-in-time, centered-in-space scheme applied to the advection equation. This scheme is

$$\frac{u_i^{n+1} - u_i^{n-1}}{2\Delta t} + c\frac{u_{i+1}^n - u_{i-1}^n}{2\Delta x} = 0, \tag{4.26}$$

which yields the recursive relation

$$u_i^{n+1} = u_i^{n-1} - \frac{c\Delta t}{\Delta x}\left(u_{i+1}^n - u_{i-1}^n\right). \tag{4.27}$$

As before, we check the stability of this scheme by substituting (4.8) into (4.27), which yields

$$
\begin{aligned}
\lambda^{n+1} A e^{\iota ki\Delta x} =& \lambda^{n-1} A e^{\iota ki\Delta x} \\
& - \frac{c\Delta t}{\Delta x}\lambda^n A \left[e^{\iota k(i+1)\Delta x} - e^{\iota k(i-1)\Delta x} \right] \\
=& \lambda^{n-1} A e^{\iota ki\Delta x} \\
& - \frac{c\Delta t}{\Delta x}\lambda^n A e^{\iota ki\Delta x} \left[e^{\iota k\Delta x} - e^{-\iota k\Delta x} \right].
\end{aligned}
\tag{4.28}
$$

Making use of (4.23) in (4.28) and rearranging gives

$$\lambda^2 + \left(2\iota \frac{c\Delta t}{\Delta x}\sin k\Delta x\right)\lambda - 1 = 0, \tag{4.29}$$

a quadratic equation for the amplification factor. Equation (4.29) has two complex roots given by

$$\lambda = \pm\sqrt{1 - \left(\frac{c\Delta t}{\Delta x}\right)^2 \sin^2 k\Delta x} - \iota\frac{c\Delta t}{\Delta x}\sin k\Delta x. \tag{4.30}$$

This means that there are two distinct solutions to the finite-difference equation, (4.26). The implications of having two solutions will be discussed later. For now, we are interested in the stability of the scheme, so we must analyze the two amplification factors to see under what conditions their magnitudes do not exceed unity. The analysis is aided if we define

$$\sigma = \frac{c\Delta t}{\Delta x} \sin k\Delta x \tag{4.31}$$

so that (4.30) can be written as

$$\lambda = \pm\sqrt{1 - \sigma^2} - \iota\,\sigma. \tag{4.32}$$

If $\sigma^2 > 1$, then (4.32) is completely imaginary and will have an amplitude given by

$$|\lambda| = \left|\pm\sqrt{\sigma^2 - 1} - \sigma\right|. \tag{4.33}$$

For this case, if $\sigma < -1$, then the positive root of (4.33) will have a magnitude greater than one, and if $\sigma > +1$, the negative root will have a magnitude greater than one. Thus the scheme is unstable for $\sigma^2 > 1$. However, if $\sigma^2 \le 1$, then (4.32) is complex and has a magnitude of

$$|\lambda| = \left|\sqrt{\left(\sqrt{1 - \sigma^2}\right)^2 + \sigma^2}\right| = 1, \tag{4.34}$$

and thus is stable. This establishes the condition for stability as

$$|\sigma| \le 1 \tag{4.35}$$

or

$$\left|\frac{c\Delta t}{\Delta x} \sin k\Delta x\right| \le 1. \tag{4.36}$$

The most limiting case is when $k = 0.5k_{\max} = \pi/(2\Delta x)$, in which case $\sin k\Delta x = 1$. Thus, to ensure stability for all wave numbers, the stability condition for the centered-in-time, centered-in-space scheme reduces to the CFL stability condition,

$$\left|\frac{c\Delta t}{\Delta x}\right| \le 1. \tag{4.37}$$

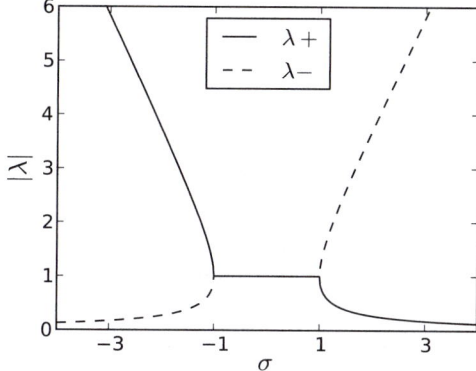

Fig. 4.7: Magnitude of the amplification factor versus σ for the centered-in-time, centered-in-space scheme. The solid line represents the positive root of (4.32), while the dashed line represents the negative root. For stability $|\sigma|$ must be less than or equal to 1.

The behavior of the amplitude factors is illustrated in Fig. 4.7, which shows a plot of the magnitude of the amplification factors versus σ for both the positive and negative roots of (4.32).

As long as the CFL criterion is met, the leapfrog scheme is neutrally stable, neither growing nor decaying with time. This is ideal because the solution to the differential equation also neither grows nor decays with time, so the numerical solution has similar behavior to the analytical solution in this regard. Fig. 4.8 shows the result of the leapfrog scheme for the same values of c, Δt, and Δx that were used in the upwind scheme of Fig. 4.3. From the figure we see that the leapfrog scheme does a much better job preserving the amplitude and shape of the solution.

4.4.7   The computational mode

The leapfrog scheme has one glaring drawback. It has two amplification factors, and therefore has two solutions! This is a problem because the advection equation only has one solution. Why does the leapfrog scheme result in two solutions? It is because it is a two-level scheme, and all two-level schemes result in two solutions. One of the solutions corresponds to the actual solution to (4.3) and is called the *physical mode*. It is this solution that contains meaning-

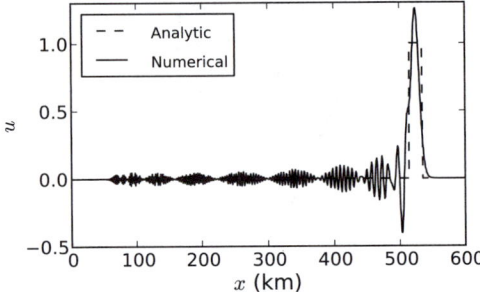

Fig. 4.8: Numerical solution to the advection equation after 500 time steps, using the *centered-in-time, centered-in-space* finite-difference scheme with the same parameters and initial conditions as in Fig. 4.3. The analytical solution is also shown. The computational mode appears as the 'noise' trailing the main signal.

ful information relative to the phenomenon being simulated. The other solution has no basis in reality. It is simply an artifact of using a multi-level scheme. It is called the *computational mode*,[4] and has two unique and troublesome characteristics: 1) it propagates in the opposite direction of the physical mode, and 2) it switches signs at every time step. The computational mode appears in Fig. 4.8 as the 'noise' trailing behind the main part of the signal.

Which root of (4.32) corresponds to the physical mode, and which corresponds to the computational mode, depends on whether c is positive or negative. The root that corresponds to the physical mode for positive values of c switches to correspond to the computational mode for negative values of c, and vice versa.

If the initial condition is fairly smooth the computational mode is not excited as much as for sharply-featured initial conditions. This is illustrated in Fig. 4.9 using an initially Gaussian initial condition. Notice that the computational mode is much more subdued compared to that of Fig. 4.8.

Despite the presence of the computational mode, the leapfrog scheme is still widely used in numerical models because it is neutrally stable, second-order accurate, and easy to program. Meth-

---

[4]In general, any multi-level scheme will have at least one computational mode. An N-level scheme will have one physical mode and $N - 1$ computational modes.

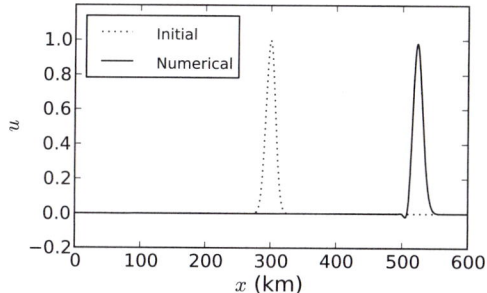

Fig. 4.9: Numerical solution to the advection equation after 500 time steps, using the *centered-in-time, centered-in-space* finite-difference scheme with the same parameters as in Fig. 4.8, only using a Gaussian, rather than box-shaped, initial condition. The computational mode is excited less when the signal is smooth.

ods for dealing with the computational mode are discussed in Section 4.5.

4.4.8 Implicit schemes

Another possible finite-difference scheme to apply to the advection equation is the backward-in-time, centered-in-space scheme. This is represented as

$$\frac{u_i^{n+1} - u_i^n}{\Delta t} + c\frac{u_{i+1}^{n+1} - u_{i-1}^{n+1}}{2\Delta x} = 0, \tag{4.38}$$

which leads to the recursive relation

$$u_i^{n+1} = u_i^n - \frac{c\Delta t}{2\Delta x}\left(u_{i+1}^{n+1} - u_{i-1}^{n+1}\right). \tag{4.39}$$

This is an example of an *implicit* scheme since the time step $n+1$ appears on both sides of the recursive relation. Before using this scheme we will check its stability. We do this as usual by substituting (4.8) into (4.39) and solving for the amplification factor, the result being

$$\lambda = \frac{1 - \iota\,(c\Delta t/\Delta x)\sin k\Delta x}{1 + (c\Delta t/\Delta x)^2\sin^2 k\Delta x}. \tag{4.40}$$

The magnitude of (4.40) is

$$|\lambda| = \frac{1}{\sqrt{1 + (c\Delta t/\Delta x)^2\sin^2 k\Delta x}}. \tag{4.41}$$

The amplification factor for this scheme is always less than unity, and therefore the scheme is always stable regardless of how long of a time increment is used. This is a significant advantage to the backward scheme. However, the solution does decay with time, with the shortest wavelengths decaying the fastest.

Though this scheme (like many implicit schemes) has the advantage of allowing use of a very long time increment, the solution is not straightforward. This is because information from the future time step appears on both sides of (4.39). If the grid consists of a total of NX grid points, there will be NX equations to solve. They will form a closed set with the variables being

$$u_0^{n+1}, u_1^{n+1}, \cdots, u_{i-1}^{n+1}, u_i^{n+1}, u_{i+1}^{n+1}, \cdots, u_{NX-2}^{n+1}, u_{NX-1}^{n+1}.$$

We can illustrate this using an example of a model with only seven grid points. We will assume fixed boundary conditions, meaning that u remains constant at the end points. Equation (4.39) is used for the interior grid points, with the boundaries remaining fixed at a constant value. The set of prediction equations is

$$
\begin{aligned}
u_0^{n+1} &= u_0^n \\
u_1^{n+1} &= u_1^n - \sigma u_2^{n+1} + \sigma u_0^{n+1} \\
u_2^{n+1} &= u_2^n - \sigma u_3^{n+1} + \sigma u_1^{n+1} \\
u_3^{n+1} &= u_3^n - \sigma u_4^{n+1} + \sigma u_2^{n+1} \\
u_4^{n+1} &= u_4^n - \sigma u_5^{n+1} + \sigma u_3^{n+1} \\
u_5^{n+1} &= u_5^n - \sigma u_6^{n+1} + \sigma u_4^{n+1} \\
u_6^{n+1} &= u_6^n
\end{aligned}
\tag{4.42}
$$

where we define the constant

$$\sigma = \frac{c \Delta t}{2 \Delta x}. \tag{4.43}$$

The system of (4.42) can be rearranged as

$$
\begin{aligned}
u_0^{n+1} &= u_0^n \\
\sigma u_0^{n+1} - u_1^{n+1} - \sigma u_2^{n+1} &= -u_1^n \\
\sigma u_1^{n+1} - u_2^{n+1} - \sigma u_3^{n+1} &= -u_2^n \\
\sigma u_2^{n+1} - u_3^{n+1} - \sigma u_4^{n+1} &= -u_3^n \qquad\qquad (4.44) \\
\sigma u_3^{n+1} - u_4^{n+1} - \sigma u_5^{n+1} &= -u_4^n \\
\sigma u_4^{n+1} - u_5^{n+1} - \sigma u_6^{n+1} &= -u_5^n \\
u_6^{n+1} &= u_6^n .
\end{aligned}
$$

In matrix form, this system of equations is

$$
\begin{pmatrix}
-1 & 0 & 0 & 0 & 0 & 0 & 0 \\
\sigma & -1 & -\sigma & 0 & 0 & 0 & 0 \\
0 & \sigma & -1 & -\sigma & 0 & 0 & 0 \\
0 & 0 & \sigma & -1 & -\sigma & 0 & 0 \\
0 & 0 & 0 & \sigma & -1 & -\sigma & 0 \\
0 & 0 & 0 & 0 & \sigma & -1 & -\sigma \\
0 & 0 & 0 & 0 & 0 & 0 & -1
\end{pmatrix}
\begin{pmatrix}
u_0^{n+1} \\ u_1^{n+1} \\ u_2^{n+1} \\ u_3^{n+1} \\ u_4^{n+1} \\ u_5^{n+1} \\ u_6^{n+1}
\end{pmatrix}
=
\begin{pmatrix}
-u_0^n \\ -u_1^n \\ -u_2^n \\ -u_3^n \\ -u_4^n \\ -u_5^n \\ -u_6^n
\end{pmatrix} .
$$

$$(4.45)$$

The coefficient matrix is a type of *sparse* matrix called a *tri-diagonal* matrix. This system of equations can be solved numerically on a computer, but it is computationally intensive. If there are 100 grid points in the model, the matrix will be 100×100. Any advantage gained by using a larger time increment may be lost by having to solve a large system of equations. However, advances in algorithms to solve systems of equations numerically have been made, and implicit schemes are now used in some models, particularly in climate models where long time increments are very advantageous.

4.5   Minimizing the Computational Mode

4.5.1   Time filtering

One method to minimize the effects of the computational mode is to apply a time filter to the solution to smooth out the short-wavelength perturbations associated with the computational mode.

A commonly used filter is the Asselin-Roberts filter. This filter is implemented by first calculating the value of u at time step $n+1$ using the leapfrog differencing scheme,

$$u_i^{n+1} = u_i^{n-1} - \frac{c\Delta t}{\Delta x}\left(u_{i+1}^n - u_{i-1}^n\right). \tag{4.46}$$

A new, filtered value for u at time step n is then obtained by using the expression

$$\tilde{u}_i^n = u_i^n + \gamma\left(u_i^{n+1} - 2u_i^n + u_i^{n-1}\right). \tag{4.47}$$

The adjustable parameter γ in (4.47) is the *filter factor*, with higher values resulting in more filtering. In practice, filter factors between 0.05 and 0.2 are common. Note that filtering is applied to time step n, after we have already calculated the value at time step $n+1$. The new filtered value, $\tilde{u}^n$, is then used in place of u^n for all subsequent calculations.

   Fig. 4.10 shows the effects of applying different filter factors to the originally rectangular disturbance used in Fig. 4.8. As shown in the bottom panel, using too large of a filter factor results in excessive damping of the solution. A filter factor of 0.2 works reasonably well, preserving the amplitude of the solution while minimizing the noise due to the computational mode.

4.5.2   Stability of the filtered scheme

One drawback of filtering is that it degrades the accuracy of the finite-difference scheme, converting the leap-frog scheme to a less-than-second-order accurate scheme. Filtering also alters the stability condition for the leapfrog scheme, with a larger filter factor placing more restrictions on the stability condition. The amplification factors for the filtered-leapfrog scheme are the roots of the cubic polynomial

$$\lambda^3 - (\gamma - 2\iota\sigma)\lambda^2 - (1 - 2\gamma)\lambda - \gamma = 0, \tag{4.48}$$

(see Exercise 4.6) where σ is defined as in (4.31). There are three roots to (4.48) because the filtered leapfrog scheme is actually a three-level scheme. Note that if $\gamma = 0$, then (4.48) is identical to (4.29).

   Figure 4.11 shows the amplification factors from (4.48) for a filter factor of $\gamma = 0.2$. The stable region for this case lies in the range

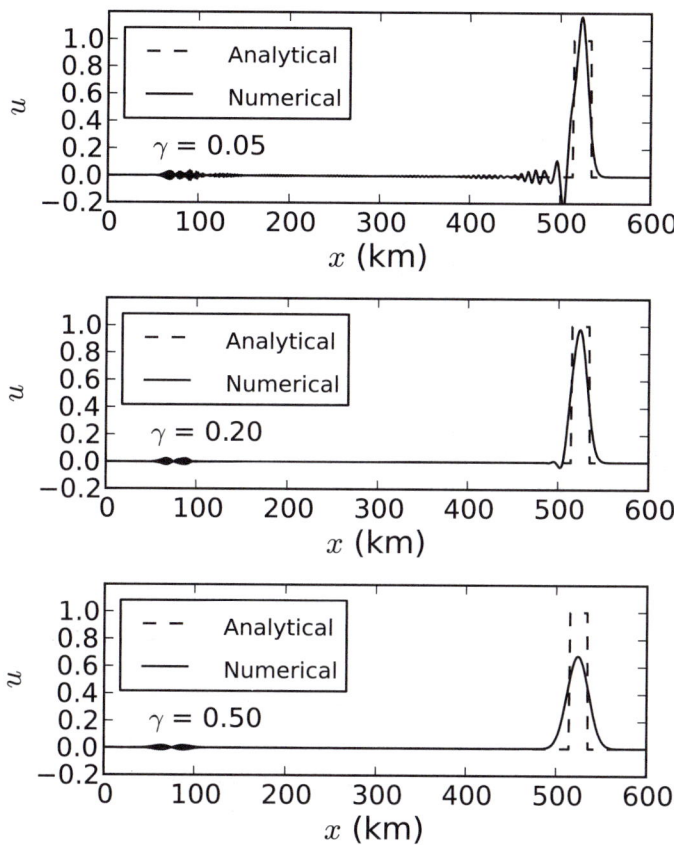

Fig. 4.10: Values of u after 500 time steps using the *centered-in-time, centered-in-space* finite-difference form of the advection equation ($c = 15$ m s^{-1}, $\Delta x = 1$ km, and $\Delta t = 30$ s) with an Asselin-Roberts time filter. The filter factors are (top to bottom) 0.05, 0.2, and 0.5 respectively.

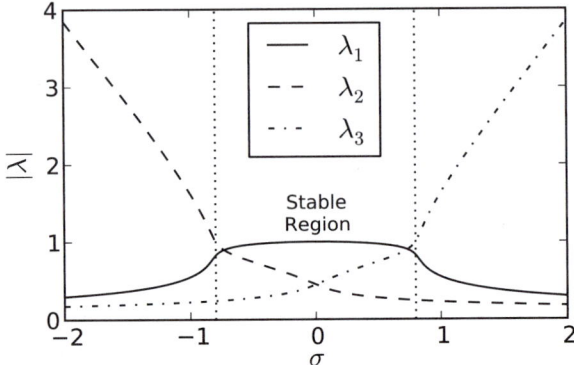

Fig. 4.11: Magnitudes of amplification factors vs. σ for the Asselin-Roberts-filtered leapfrog scheme for the 1D advection equation. The amplification factors are the roots of (4.48) using a filter factor $\gamma = 0.2$. The physical mode is λ_1. The other two modes are computational modes. Dotted vertical lines denote $\sigma = -0.8$ and 0.8.

$-0.8 \leq \sigma \leq 0.8$. The greater the filter factor, the more stringent is the stability condition.

4.5.3   Other methods

In addition to filtering, there are also other, more complicated methods to minimize the effects of the computational mode. One such method is to periodically substitute a different scheme in place of the leapfrog scheme. The substituted scheme can be either a single-level scheme or a highly-damped scheme to eliminate or suppress the computational mode. The Matsuno scheme (see Appendix B, Section B.3.2) is often employed for such a role. As soon as the leapfrog scheme is restarted, the computational mode will again appear, but hopefully at reduced amplitude.

4.6   The Adams-Bashforth Scheme

Of the three time-differencing schemes we have applied so far to the advection equation, the most suitable seems to be the leapfrog scheme, since it has an amplification factor of one as long as the CFL condition is met. However, it has the disadvantage of possessing a computational mode. There are numerous higher-order time schemes that we could try as well. These higher-order schemes have the drawback that, though they are more accurate, they also require many more floating-point operations to compute, which means they take longer and require more memory. However, as computers have become faster and have more memory, these higher-order schemes are more practical to apply. One scheme that seems to strike an appropriate balance between accuracy and quickness is the third-order Adams-Bashforth scheme. Using the general notation from Chapter 3 for an equation of the form

$$\frac{dF}{dt} = G(F), \tag{4.49}$$

the Adams-Bashforth scheme has the form

$$F^{n+1} = F^n + \frac{\Delta t}{12}\left(23G^n - 16G^{n-1} + 5G^{n-2}\right). \tag{4.50}$$

When applied to the advection equation with centered-in-space differencing the resulting recursive equation is

$$\begin{aligned} u_i^{n+1} = u_i^n &- \frac{c\Delta t}{24\Delta x}\Big[23\left(u_{i+1}^n - u_{i-1}^n\right) \\ &- 16\left(u_{i+1}^{n-1} - u_{i-1}^{n-1}\right) + 5\left(u_{i+1}^{n-2} - u_{i-1}^{n-2}\right)\Big]. \end{aligned} \tag{4.51}$$

Although this scheme actually has two computational modes, both computational modes are highly damped and do not require filtering or other special methods to reduce their influence. Fig. 4.12 shows the results of using the 3rd-order Adams-Bashforth scheme for the advection equation. The solution does a good job of maintaining the amplitude of the solution without much noise from the computational modes.

The amplification factor for this scheme is given by a cubic polynomial (see Exercise 4.5),

$$\lambda^3 - \left[1 - \iota\left(\frac{23}{12}\right)\sigma\right]\lambda^2 - \left[\iota\left(\frac{4}{3}\right)\sigma\right]\lambda + \iota\left(\frac{5}{12}\right)\sigma = 0, \tag{4.52}$$

where σ is again defined as in (4.31). Equation (4.52) has three roots, corresponding to the single physical mode and two computational modes. Fig. 4.13 shows a plot of the magnitudes of the three roots of (4.52) as a function of σ. This plot shows that for the 3rd-order

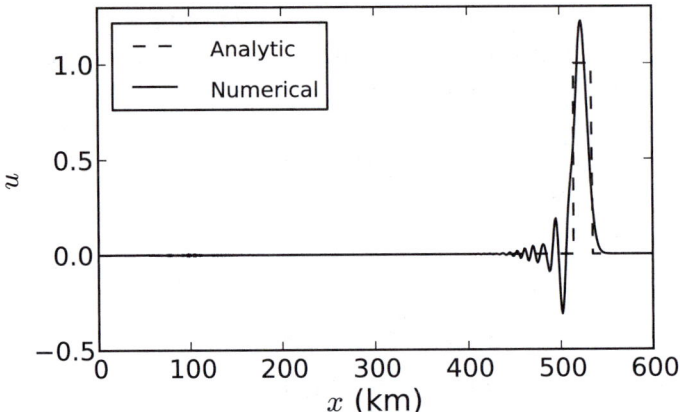

Fig. 4.12: Solution after 500 time steps (bottom) using the 3rd-order Adams-Bashforth time scheme with centered-in-space finite-differencing for the advection equation ($c = 15$ m s^{-1}, $\Delta x = 1$ km, and $\Delta t = 30$ s). Dashed line is analytic solution.

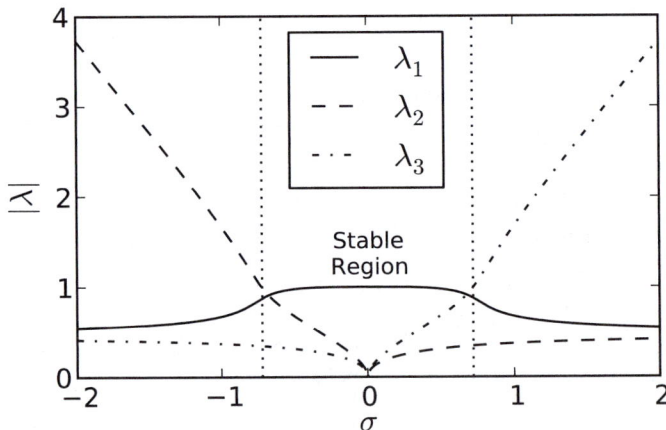

Fig. 4.13: Magnitudes of amplification factors vs. σ for the 3rd-order Adams-Bashforth scheme for the 1D advection equation. The amplification factors are the roots of (4.52). The solid curve is the physical mode, while the other two curves are the computational modes. Dotted vertical lines denote $\sigma = -0.72$ and 0.72.

Adams-Bashforth scheme to remain stable, $|\sigma|$ must be less than 0.72, a more restrictive condition than for the unfiltered leapfrog scheme. Outside of this range, one of the computational modes has an amplitude exceeding one, resulting in instability.

4.7 The Diffusion Equation

Another equation that appears frequently in atmospheric models is the diffusion equation, which in one-dimension is

$$\frac{\partial u}{\partial t} = K\frac{\partial^2 u}{\partial x^2}. \tag{4.53}$$

This equation appears when the simulation must account for either molecular or turbulent diffusion of momentum, heat, moisture, aerosols, or other atmospheric trace gases. The parameter K is the *diffusion coefficient*.

The stability criterion for the diffusion equation is quite different than that for the advection equation. In fact, the leapfrog scheme, which was stable for the advection equation as long as the CFL stability condition was met, is *always unstable* when applied to the diffusion equation. The forward scheme is stable for the diffusion equation under certain conditions. The amplification factor for the forward scheme is (see Ex. 4.3)

$$\lambda = 1 - 2\frac{K\Delta t}{\Delta x^2}(1 - \cos k\Delta x). \tag{4.54}$$

This will be stable as long as

$$K\Delta t/\Delta x^2 \leq 1/2.$$

However, for

$$1/4 < K\Delta t/\Delta x^2 \leq 1/2$$

the amplification factor from (4.54) may be negative, which is undesirable since the solution would flip signs at each time step. Therefore, when using the forward-in-time scheme for the diffusion equation, the time step and grid spacing should be chosen such that

$$K\Delta t/\Delta x^2 \leq 1/4. \tag{4.55}$$

4.8   Operator Splitting

Atmospheric numerical models need to solve complex equations that involve a combination of terms. The stability criteria for a finite-difference scheme may be different for each term of the equation being solved. A scheme that is stable for one term may be unstable for another. As an example, consider a linear, one-dimensional equation that includes both advection and diffusion terms,

$$\frac{\partial u}{\partial t} = -c\frac{\partial u}{\partial x} + K\frac{\partial^2 u}{\partial x^2}. \tag{4.56}$$

We would like to use leapfrog time differencing to take advantage of its second-order accuracy, but the leapfrog scheme is unstable with regard to the diffusion term. However, through the concept of *operator splitting*, we can still use the leapfrog scheme when calculating the effect of the advection term but then use the forward scheme, which is stable under the right circumstances when applied to the diffusion term, to calculate the effects of diffusion. In fact, we can even use a different time increment for each term, perhaps calculating advection every time step, but only calculating diffusion every fourth time step.[5]

To illustrate operator splitting, imagine that we are building a simple model that will simulate the one-dimensional advection and dispersion of a non-reactive smoke plume. In this case, (4.56) contains the entire physics of the problem to be solved, since the only two processes affecting plume concentration are advection and dispersion.[6] In our simulation, u will represent the concentration of the plume, while c is the mean wind speed. We want to use a grid spacing of 1 km and we expect that the maximum wind speed for our simulation will not exceed 25 m s^{-1}. This requires an advective time increment of less than 40 seconds in order to meet the CFL stability condition. To allow a margin of error, we choose 30 seconds. For the diffusion term, we assume a typical eddy diffusion

---

[5]In general, *operator splitting* refers to treating different terms in an equation with different finite-differencing operators. If different time increments are also used for each term, it is also referred to as *time splitting*.

[6]If the plume were reactive, an additional term or terms representing loss or addition to the plume via chemical reactions or deposition to the ground would be needed.

coefficient of 500 m^2 s^{-1}, so that the maximum diffusion time increment is 1000 seconds. Even though we could get away with using a diffusion time increment this large and still have stability, a smaller time increment is expected to yield more accurate results. We therefore might choose to calculate diffusion every 120 seconds. If Δt represents the advective time increment, then the diffusion time increment is $4\Delta t$.

When the model is running, it will calculate the new plume concentration every time step using leapfrog time differencing via the following equation,

$$u_i^{n+1} = u_i^{n-1} - \frac{c\Delta t}{\Delta x}\left(u_{i+1}^n - u_{i-1}^n\right), \tag{4.57}$$

which accounts for advection only. However, at every fourth time step the diffusion term is calculated using the forward scheme with a time interval of $4\Delta t$. So, while (4.57) is used as the prediction equation for three out of four time steps, for the fourth time step the equation

$$\begin{aligned} u_i^{n+1} = &u_i^{n-1} - \frac{c\Delta t}{\Delta x}\left(u_{i+1}^n - u_{i-1}^n\right) \\ &+ \frac{K\left(4\Delta t\right)}{\Delta x^2}\left(u_{i+1}^{n-3} + u_{i-1}^{n-3} - 2u_i^{n-3}\right) \end{aligned} \tag{4.58}$$

is used. Operator splitting allows the simulation to complete much more quickly because there are fewer computations required to advance the solution.

Operator splitting can also be used when extending a one-dimensional scheme to two dimensions. For example, the two-dimensional advection equation could be split by first computing and applying the x-direction advection operator, followed by applying the y-dimension advection operator. Such splitting is only necessary for more complicated schemes, and is useful because an operator split scheme is often more computationally efficient than an unsplit scheme. However, operator splitting also introduces additional errors into the computation. These errors are discussed further in Appendix D.

4.9   The CFL Condition in Two Dimensions

The advection equation in two dimensions is

$$\frac{\partial u}{\partial t} + c_x \frac{\partial u}{\partial x} + c_y \frac{\partial u}{\partial y} = 0, \tag{4.59}$$

where c_x and c_y are the advection velocity components in the x and y directions. The total advection speed is

$$c = \sqrt{c_x^2 + c_y^2}. \tag{4.60}$$

The stability conditions for two dimensional grids are slightly more stringent than for one-dimensional grids. This is because a two-dimensional grid having a grid spacing of d in both the x and y directions has an *effective grid spacing* in the diagonal direction of

$$d' = d/\sqrt{2}. \tag{4.61}$$

This is illustrated in 4.14, which shows the crests of the shortest resolvable sinusoidal wave traveling purely in the x direction and also traveling diagonally. In order to resolve a sinusoidal feature there must be a minimum of three independent grid points within one wavelength. In the purely x direction, these three grid points have a spacing of d, but diagonally they have a spacing of $d/\sqrt{2}$, so shorter features can be resolved in the diagonal direction. This shorter grid spacing requires a shorter time interval in order to meet stability requirements.

Based on this analysis we can find the stability conditions in two dimensions for any of the schemes we have already considered in one dimension. We merely need to replace Δx with d'. So for the unfiltered leapfrog scheme applied to the advection equation in 2D, the stability condition is

$$\left| \frac{c \Delta t}{d'} \right| \leq 1 \tag{4.62}$$

or

$$\left| \frac{c \Delta t}{d} \right| \leq \frac{\sqrt{2}}{2}. \tag{4.63}$$

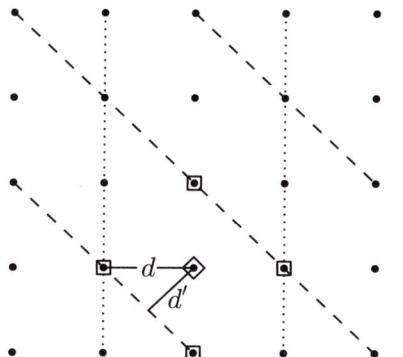

Fig. 4.14: A regular 2D grid with grid spacing d. The dotted lines show wave crests for the shortest resolvable wave traveling purely in the x direction. The dashed lines show the crests of the shortest resolvable wave traveling diagonally. The four grid points annotated with squares indicate the four grid points used for the centered-in-space finite-difference representation of the 2D advection equation at the diamond-annotated grid point. The *effective grid spacing* for waves traveling diagonally is $d' = d/\sqrt{2}$.

Exercises

Ex. 4.1: Substitute (4.8) into (4.16) and solve for the magnitude of the amplification factor to verify that the stability condition for the forward-in-time, forward-in-space scheme applied to the 1D advection equation is

$$-1 \leq c\Delta t/\Delta x \leq 0.$$

Ex. 4.2: Substitute (4.8) into (4.39) and solve for the magnitude of the amplification factor to verify that the backward-in-time, centered-in-space scheme applied to the 1D advection equation is always stable.

Ex. 4.3:
a. Show that the amplification factor for the forward-in-time scheme applied to the diffusion equation is

$$\lambda = 1 - 2\mu \left(1 - \cos k\Delta x\right),$$

where $\mu = K\Delta t/\Delta x^2$.

b. From the result above, show that this scheme is stable as long as $\mu \leq 1/2$.

c. Show that a nonnegative, stable amplification factor is guaranteed only for $\mu \leq 1/4$.

Ex. 4.4:   Show that the amplification factor for the leapfrog scheme applied to the diffusion equation is

$$\lambda = -2\mu \left(1 - \cos k\Delta x\right) \pm \sqrt{4\mu^2 \left(1 - \cos k\Delta x\right)^2 + 1},$$

where $\mu = K\Delta t / \Delta x^2$. Also show that one of the roots always has a magnitude greater than one, and so this scheme is always unstable.

Ex. 4.5:

a. Substitute (4.8) into (4.51) to show that the amplification factor for the 3rd-order Adams-Bashforth scheme applied to the 1D advection equation satisfies the cubic polynomial

$$\lambda^3 - \left[1 - \iota \left(\frac{23c\Delta t}{12\Delta x}\right) \sin k\Delta x\right] \lambda^2 - \left[\iota \left(\frac{4c\Delta t}{3\Delta x}\right) \sin k\Delta x\right] \lambda$$
$$+ \iota \left(\frac{5c\Delta t}{12\Delta x}\right) \sin k\Delta x = 0.$$

b. Show that when $k = k_{\max}$ one amplification factor has a value of one, while the other two are zero.

Ex. 4.6:

The Asselin-Roberts-filtered leapfrog scheme can also be written as

$$u_i^{n+1} = \tilde{u}_i^{n-1} - \frac{c\Delta t}{\Delta x} \left(u_{i+1}^n - u_{i-1}^n\right) \tag{4.64}$$

where

$$\tilde{u}_i^{n-1} = u_i^{n-1} + \gamma \left(u_i^n - 2u_i^{n-1} + u_i^{n-2}\right). \tag{4.65}$$

These two equations can be merged into a single equation by substituting (4.65) into (4.64). Merge the two equations, and assess the stability of the scheme by substituting (4.8) into the resulting equation, deriving (4.48).

## CHAPTER 5

Programming Numerical Models

5.1 Overview

Now that we understand the fundamentals of finite differencing and how to apply it to simple partial differential equations, it is time to discuss how these solutions are implemented in a computer program. Programs for numerically solving finite-difference equations can be written in any computer language. Compiled languages such as Fortran, C, and C++ are generally faster and more efficient than interpreted languages, and therefore operational models are generally programmed using a compiled language. However, models can just as easily be written in interpreted[1] languages such as Perl,

---

[1]Compilers read the entire source code, written in a text file, and convert it into an executable code (machine language), often optimized, that can be directly executed by the operating system. Interpreters either read the text file line-by-line and convert each statement into machine language and execute it as it is read, or convert the text-file statements into an intermediate code that is then further converted into machine language for execution by the operating system. Compiled languages are generally more efficient and faster than interpreted languages, since the conversion to machine language only occurs one time, and the machine language code can be executed whenever the program needs to be run. In an interpreted language, the interpretation process occurs every time the program is run. However, interpreted languages are often more flexible and changes can be easily made to the program at runtime.

Python, Ruby, etc.

Although considered obsolete by computer scientists, Fortran is still the primary language for large-scale fluid computation applications. This is because much effort and many resources have been applied to optimizing Fortran compilers for high-speed, parallel computation. Fortran's dominance in this area does not seem to be in danger, and it will likely remain in either new or legacy applications for many decades.

A complex operational model would likely be written in a compiled language such as Fortan, but there are many other languages that can be used for smaller-scale research or teaching models. There are several proprietary languages, such as IDL and MATLAB, that have built-in graphics capabilities which allow the results of the model to be displayed as the model is running. A language that is seeing growth in the atmospheric and ocean sciences communities is Python, particularly when coupled with the MATPLOTLIB plotting library or some other library for creating plots and graphs. One big advantage of Python over IDL and MATLAB is that Python does not require the purchase of a license and is easily and freely available.

This chapter is not a primer in programming with a particular language. The concepts discussed here are universal. However, where specific examples are used they will be illustrated using the Python programming language. This choice is made because the syntax of Python is fairly intuitive and it is free and readily available. Keep in mind that the program examples could just as easily be written in Fortran, C, C++, Java, Ruby, Perl, IDL, MATLAB, or any other programming language.

In the rest of this chapter we present a general outline for how to write a computer program to solve a partial differential equation, such as the advection equation. We then present a specific example of a program to solve the 1D advection equation using the leapfrog time scheme and centered space differencing. This chapter concludes with a discussion of programming strategies and tips for writing programs that will execute efficiently and quickly.

5.2   General Outline of an Algorithm

Regardless of the partial differential equation to be solved numerically, there are certain general steps that must be performed. These are:

Step 1 –   Declare, if necessary, and initialize any variables and arrays.

Step 2 –   Open any input and/or output files.

Step 3 –   Initialize the solution for time step zero.

Step 4 –   Enter the main loop of the program, which includes the following substeps:

>*Substep 4.1 –*   Solve the recursive equation at each grid point.
>
>*Substep 4.2 –*   Apply boundary conditions.
>
>*Substep 4.3 –*   Apply any necessary filtering.
>
>*Substep 4.4 –*   Swap variables/arrays.
>
>*Substep 4.5 –*   Write output to a file or display device.

Step 5 –   Close input/output files.

A detailed discussion of each of these steps follows.

Step 1 – Declare, if necessary, and initialize variables and arrays. In this step, variables and arrays are created and some are also assigned initial values. The variables hold physical parameters needed for the solution of the problem. For the 1D advection equation, the advection speed, c, is one such parameter. For other model equations, physical parameters could include gravity, fluid depth, molecular or turbulent diffusivity, and others. Variables are also needed to hold model parameters such as the grid spacing, the time interval, the total number of time steps for the integration, etc.

Arrays are needed to hold the values of the solution at each time step. For a one-dimensional problem these arrays will be of length NX, which is the number of grid points in the domain. For a two-dimensional problem, a two-dimensional array of size (NX, NY) will be used, where the first is the number of grid points in the x-direction and the second is the number of grid points in the y-direction.

For the forward-in-time scheme two distinct arrays are needed for the solution. One array is for the solution at the current time step (which we will call *u_now*), and the other is for the solution at the future time step (which we will call *u_next*). For the leapfrog scheme, in which data at the past time step is used, we also need an additional array to hold the previous solution (which we will call *u_past*).

Whether or not variables need to be explicitly declared at the beginning of the model depends on whether the programming language being used is *statically-typed* or *dynamically-typed*.[2] Fortran is a statically-typed language, and therefore requires that variables be explicitly declared as a certain type (real, integer, string, etc.) prior to their use.[3] Dynamically-typed languages such as Python, do not require explicit declarations of variable type.

Step 2 – Open any input and/or output files. Often the initial conditions will be stored in a file that must be read into the program. Also, the output from the model may need to be saved for later analysis. Input and output files may be in either binary or ASCII format. ASCII has the advantage that the contents of the file can be opened and readily viewed with a text editor. However, this comes at the expense of storage space. For large-scale and operational models binary files are usually used. For smaller-scale, simple models in an instructional setting, ASCII is appropriate since the text output is readily accessible and speed and storage are not usually at a premium.

Step 3 – Initialize the solution for time step zero. The initial conditions can either be read from an external file or generated within the program. **An important caution: Never assume that the inter-**

---

[2]The terms *statically-typed* and *dynamically-typed* refer to whether or not a language requires the type of variable (real, integer, character, complex, etc.) to be explicitly declared prior to assigning the variable a value. In statically-typed languages variables must first be declared as a specific type, and once declared cannot change type during program execution. In dynamically-typed languages the type of variable is determined by the value assigned to it, and the type can change during execution by simply assigning a value of a different type to the variable.

[3]Many Fortran compilers implicitly type variables starting with the letters 'i' through 'n' as integer variables, and all others as real, unless the user specifically declares them otherwise. However, when writing a program in Fortran it is always best to use the Fortran statement 'IMPLICIT NONE', which forces the programmer to explicitly declare all variables. This saves many debugging headaches.

preter or compiler will automatically set the value of uninitialized arrays and variables to zero! Many compilers and interpreters will assign completely random numbers into uninitialized variables and arrays, and this can be a very difficult problem to detect and debug. If a certain variable or array needs to start out with a value of zero, you should always explicitly assign the value of zero to it at the beginning of the program.

Step 4 – Enter the main loop of the program. The main loop will be a loop through the time steps beginning at $n = 0$ and running to time step $n\_total - 1$, where n_total is the time step at the completion of the simulation (and is also the total number of time iterations required.) For example, if a time increment of 30 seconds is used, and the simulation is to be run for a total simulation time of 1 hour, then n_total would equal 120. Notice that the loop starts at $n = 0$ and ends at $n = n\_total - 1$, so that the contents of the loop are executed n_total number of times.

The main loop tends to be the most complicated part of the program. Within the main loop there are several substeps that must be accomplished. These include: 1) solving the recursive equation for the partial differential equation at each of the grid points in the domain, 2) applying any necessary boundary conditions, 3) filtering to reduce the computational mode, 4) swapping the variables/arrays between time steps, and 5) writing the output of the solution to a display device or to a file. These substeps are described below.

Substep 4.1 – Solve the recursive equation at each grid point. Each time through the main loop the program must solve the recursive equation appropriate to the problem to calculate the value of the solution at the next time step, $n + 1$. This means that nested within the main loop there will be an additional loop that traverses each grid point, in which the recursive relation at that grid point will be applied. This nested loop, referred to as the *grid loop*, will need to loop through the grid indexes starting at $i = 0$ and ending at $i = NX - 1$. However, for most problems the end grid points (boundaries) are treated specially. This is because usually the recursive equation used for the interior points is not appropriate at the boundaries. Thus, it is far more common for the grid loop to iterate over only the interior grid points of the domain, from $i = 1$ to $i = NX - 2$.

For the 1D advection equation using leapfrog time differencing, the form of the recursive equation inside the loop would be of the form

$$u\_next[i] = u\_past[i] - \frac{c\Delta t}{\Delta x}(u\_now[i+1] - u\_now[i-1]). \quad (5.1)$$

For a two-dimensional problem the grid loop will actually consist of two nested loops. The outer of these loops will iterate over the y-domain while the inner will iterate over the x-domain. A three-dimensional problem would have three nested grid loops.

Substep 4.2 – Apply boundary conditions. In this step the solution at the boundaries is updated. A recursive equation different from that used in the interior of the grid is usually required here. For example, (5.1) cannot be evaluated on the boundaries, where $i = 0$ or $i = NX - 1$, because the values of $u\_now[-1]$ and $u\_now[NX]$ are not defined. The simplest, though not always best, solution is to use a *fixed* boundary condition where the solution at the boundary points is constant with time. The recursive equations at the boundaries are then

$$u\_next[0] = u\_now[0]$$
$$u\_next[NX-1] = u\_now[NX-1]. \quad (5.2)$$

More complex boundary conditions are usually used in actual models. These are discussed in detail in Chapter 9.

Substep 4.3 – Apply any necessary filtering. This step is only necessary when using a scheme that has an undamped computational mode, such as the leapfrog scheme. **An important caution here is to wait until the new value of the solution is predicted for every grid point (including the boundaries) before applying the filter.** In other words, the filtering should not be done inside of the grid loop from Substep 4.1, but instead should be applied in a separate loop iterating over the grid. This avoids using a mixture of filtered and unfiltered values in the space differencing scheme of Substep 4.1. Failure to follow this caution may result in unexpected and erroneous behavior of the solution.

Substep 4.4 – Swap variables/arrays. The solutions at the different time steps are stored in separate arrays (u_past, u_now, u_next). After each iteration of the time loop we need to place the values contained in *u_next* into *u_now*. This is known as swapping the variables. If a two-level time scheme, such as the leapfrog scheme, is used, then we also need to replace the solution at the previous time step (*u_past*) with that from the current time step (*u_now*). The correct order for swapping the variables is

$$u\_past = u\_now,$$
$$u\_now = u\_next. \tag{5.3}$$

These variables must be swapped in the order shown above. If the order is reversed, each variable will end up containing the same value.

It is possible to avoid variable swapping if the solutions are stored in a multi-dimensional array where one of the indexes corresponds to the time step. However, such arrays will be exceedingly large and consume much memory. Therefore, this method should be avoided in most circumstances.

Substep 4.5 – Write output to a file or display device. Periodically the output from the model will be saved to a file for later analysis and/or display on a graphics device. It is impractical to save the data from each and every time step as the files would be very large. In practice the output is saved or displayed after some predetermined number of time steps. If this increment is denoted as *out_step*, then an *if-then* statement used in conjunction with the modulo function[4] can be used to trigger the output.

As an example, suppose we are running a model with a time increment of 30 seconds, and we wish to save the output from the model at 10 minute intervals. The parameter *out_step* would be set to a value of 20. At the conclusion of each time step the quantity '$n + 1$ modulo *out_step*' would be evaluated. If the result is zero, then $n + 1$ is an integral multiple of *out_step*, and the appropriate statements to write the data to a file or display on a graphics device can be executed.[5]

---

[4]The modulo function of two integers, a and b, returns the remainder of a/b. If a modulo b is zero, then a is an integer multiple of b.

[5]The modulo of $n + 1$ and *out_step* is used, and not n, because after the time step is completed and the variables are swapped, the variable *u_now* is valid at time step $n + 1$. If the output is accomplished at the beginning of the time step, then it would be appropriate to use 'n modulo *out_step*'.

Step 5 – Close input/output files. This is a 'housekeeping' step to ensure that input and output files are closed gracefully prior to the termination of the program.

5.3   An Example Program

In this section we show and describe the Python code for a model that solves the 1D advection equation using the forward-in-time, backward-in-space scheme for a positive value of c (this is the upwind scheme). The model uses fixed boundary conditions, a grid spacing of 1 km, 601 grid points, and a time increment of 30 seconds. The wave speed c is set at 15 m/s. The output is written to a text file every 5 minutes. Comments are indicated with a '#' character. The sections of the code are labeled to correspond to each of the steps discussed previously. The entire Python program for the completed example is shown below.

```
#----------- Python Program for Upwind Scheme -----------
# Beginning of program

import numpy as np  #  imports Numerical Python library

#-----------------------------------------------------
# Steps 1 and 2: Set up variables and arrays and
# initialize data

# The following are all physical or model parameters
nx = 601         # number of grid points
dx = 1000.0      # grid spacing
dt = 30.0        # time interval
n_total = 500    # final time step
out_step = 10    # number of time steps between output
c = 15.0         # propagation speed

# Set up arrays
u_now = np.zeros(nx)    # NumPy array initialized to zero
u_next = np.zeros(nx)   # NumPy array initialized to zero

#-----------------------------------------------------
```

```
# Step 2:  Open data file for writing model output
outfile = open('model_output.dat', 'w')


#-----------------------------------------------------
# Step 3:  Initialize solution for time level zero
u_now[nx/2-10:nx/2+10] = 1.0  # box signal of width 20


#-------------------------------------
# Step 4:  This is the time loop
for n in range(0,n_total):   # Start of time loop

# Substep 4.1: This is the grid loop for the recursive
# equation
  for i in range(1,nx-1):  #  Start of grid loop
    u_next[i]=u_now[i] - (c*dt/dx)*(u_now[i]-u_now[i-1])

  # Substep 4.2: Apply Boundary conditions
  u_next[0] = u_now[0]
  u_next[nx-1] = u_now[nx-1]

  # Substep 4.3: Filtering (not used in the upwind scheme)

  # Substep 4.4: Swap variables
  u_now = np.copy(u_next)

  # Substep 4.5: Write output if n+1 is a multiple
  # of out_step. Each record is one time level, with grid
  # values separated by commas.

  if (n+1) % out_step == 0:  # Modulo operator
    u_now.tofile(outfile, sep = ',', format = '%f')
    outfile.write('\n')    # Adds newline character
                           # at end of record
#-------------------------
# Step 5:  Close data files
outfile.close()  #  Close data file


# End of program
```

The equivalent Fortran program (without comments) is shown below.

```fortran
!-------- Fortran Program for Upwind Scheme -------------
PROGRAM upwind

IMPLICIT NONE

REAL, ALLOCATABLE :: u_next(:), u_now(:)
REAL, PARAMETER :: c = 15.0, dt = 30.0, dx = 1000.0
INTEGER :: i, n
INTEGER, PARAMETER :: nx = 601, n_total = 500, &
    width = 20, out_step = 10

ALLOCATE (u_next(nx), u_now(nx))

OPEN(20, file = 'model_output.dat', status = 'replace', &
    action = 'write')

u_now = 0.0
u_now(nx/2-width/2:nx/2+width/2) = 1.0

DO n = 0, n_total-1

 DO i = 2, nx-1
  u_next(i) = u_now(i) - (c*dt/dx)*(u_now(i)-u_now(i-1))
 END DO

 u_next(1) = u_now(1)
 u_next(nx) = u_now(nx)

 u_now = u_next

 IF (MOD(n+1,out_step) .eq. 0) WRITE(20,*) u_now

END DO

CLOSE(20)
END PROGRAM upwind
```

5.4   Programming and Optimization Tips

The examples of the previous section are not the most efficient or speediest algorithms that can be written to solve the advection equation. There are many changes we can make to speed up the execution of the program. For our small example speed is not an issue, but for a large-scale model with complex physics and many more lines of code, efficiency and speed are important. However, each compiler/interpreter is different, and algorithms that run fast on one particular compiler may be significantly slower on another. Also, modern compilers and interpreters have been optimized for efficiency and are more forgiving when it comes to inefficient program structure.

The tips presented below are generally applicable to most programming languages and compilers/interpreters.

5.4.1   Beware of integer division

Most programming languages evaluating division between two integers return an integer result. For example, the expression 4/3 would return a value of '0'. This can cause unexpected results for the unwary. It is always a good idea to use floating-point numbers when doing division. If the desired result is '1.333333', then the expression should include at least one floating-point value (e.g., 4/3.0, or 4.0/3, or 4.0/3.0).

5.4.2   Avoid repeated calculations

In many instances the same calculation occurs in multiple places within the code. For example, the combination $c\Delta t/\Delta x$ occurs in numerous locations inside a typical model. Rather than repeating this calculation multiple times, it is better to define a new variable equal to this result. Then, whenever the combination $c\Delta t/\Delta x$ is needed, the new variable is substituted, rather than performing the calculation. An example of this occurs inside the grid loop for the 1D advection model, where quantity $c\Delta t/\Delta x$ is needed at each grid point. In order to keep from having to compute this at every grid point, we instead define it outside of the loop. The code for Substep 4.1 would then look like

```
sigma = c*dt/dx
for n in range(0,n_total):
    for i in range(1, nx-1):
        u_next[i] = u_now[i] -sigma*(u_now[i]-u_now[i-1])
```

In a test comparing the average CPU run time of the upwind model
for the 1D advection equation using the original code from the prior
section, and with the change above, the small change resulted in an
average six percent faster execution time over 100 model runs.

5.4.3   Use implied loops

Even greater execution speed is achieved by the use of implied
loops (also called implicit loops.) As an example, we could have
written the grid point loop, which in our Python example is

```
for i in range(1,nx-1):
    u_next[i]=u_now[i] - sigma*(u_now[i]-u_now[i-1])
```

as a single line of code,[6]

```
u_next[1:nx-1]=u_now[1:nx-1]-sigma*(u_now[1:nx-1]- \
    u_now[0:nx-2])
```

(the '\' character in Python indicates a continuation line.) In a
benchmark test the version of the 1D model using the implied loop
executed 24 times faster than the version with the explicit loop.
Therefore, implicit loops should always be used in numerical models when possible.

5.4.4   Move IF statements outside of loops

It takes time for a program to evaluate an IF statement. Placing
an IF statement outside of a loop can therefore save time during
program execution. A common example of this arises when solving
the advection equation using the leapfrog scheme. For the first pass
through the time loop we need to use a forward-in-time scheme,
such as the upwind scheme, since we do not have information for
the past time step. We could write the program as:

---

[6]Note that in Python an index range expressed as [1:n] runs from indices 1 to
$n - 1$, and excludes the last index n. This is a quirk of Python. In most other
languages the index range includes the last index.

```
for n in range(0,n_total):
  if (n == 0):
    u_next[1:nx-1]=u_now[1:nx-1]-sigma*(u_now[1:nx-1]- \
      u_now[0:nx-2])
  else:
    u_next[1:nx-1]=u_past[1:nx-1]-sigma*(u_now[2:nx]- \
      u_now[0:nx-2])
```

. . . [other code for boundaries, filtering, etc] . . .

but this requires an if-else construct within the time loop. Another method is to do the forward scheme for the first time step outside of the loop and then enter the loop starting with the second time step. This would look like

```
u_next[1:nx-1]=u_now[1:nx-1]-sigma*(u_now[1:nx-1]- \
  u_now[0:nx-2])
```

. . . [other code for boundaries, filtering, variable swapping, etc.] . . .

```
for n in range(1,n_total):
  u_next[1:nx-1] = u_past[1:nx-1]-sigma*(u_now[2:nx]- \
    u_now[0:nx-2])
```

. . . [other code for boundaries, filtering, variable swapping, etc.] . . .

Note that when entering the loop, the index starts at one, not zero as before.

5.4.5 Avoid unnecessary print/write statements

Input and output operations are also time consuming, and should only be used where necessary.

Exercises

Ex. 5.1:   Either copy, or use as a guide, the example program from this chapter to write a program to solve the 1D advection equation using the forward-in-time, backward-in-space scheme on a grid containing 601 grid points. Use a wave speed of 15 m s^{-1}, a grid spacing of 1 km, and a time increment of 30 s. Initialize the solution with a box of width 20 grid points, centered on the grid. The results after 500 iterations should look like the numerical solution in Fig. 4.3.

Ex. 5.2:   Experiment with a copy of the program from the previous problem, by using different time steps and wave speeds. Specifically:
a. Use a negative wave speed to convert the scheme to a downwind scheme, and see that it is unstable.
b. Use a large, positive wave speed that exceeds the CFL condition to see that it is unstable.

Ex. 5.3:   Modify a copy of the program from Ex. 5.1 to convert it into a forward-in-time, forward-in-space scheme. Use the same parameters as in Ex. 5.1, only with a negative wave speed. The results after 500 iterations should look similar to the numerical solution in Fig. 4.3, except the solution would have moved to the left instead of to the right.

Ex. 5.4:   Modify a copy of the program from Ex. 5.3. Use the exact same parameters as in Ex. 5.3, only with a positive wave speed, to see that it is unstable.

Ex. 5.5:   Write a program to solve the 1D advection equation using the centered-in-time (leapfrog), centered-in-space scheme with the same parameters and initial conditions as in Ex. 5.1. The results after 500 iterations should look like the numerical solution in Fig. 4.8.

Ex. 5.6:   Modify a copy of your program from Ex. 5.5 to add an Asselin-Roberts time filter. Run the program for filter factor values of 0.05, 0.2, and 0.5, and check the results with those of Fig. 4.10.

Ex. 5.7:   Write a program to solve the 1D advection equation using the 3rd-order Adams-Bashforth scheme with the same parameters and initial conditions as in Ex. 5.1. The solution after 500 iterations should look similar to Fig. 4.12.

Ex. 5.8:   Write a program to solve the 1D diffusion equation using the forward-in-time scheme. Use a grid with 601 points and a grid spacing of 1 km, a diffusion coefficient of 4000 m^2 s^{-1}, and a time interval of 120 s. Initialize the solution with a box of width 20 grid points, centered on the grid. After 125 iterations your results should look like the figure below.

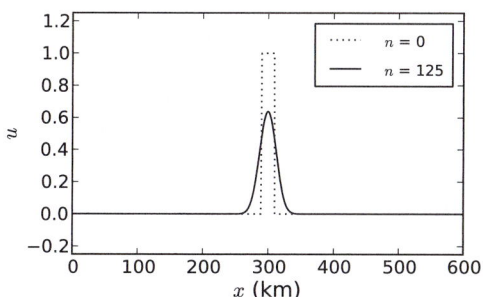

Ex. 5.9: Write a program to solve the 1D combined advection-diffusion equation

$$\frac{\partial u}{\partial t} = -c\frac{\partial u}{\partial x} + K\frac{\partial^2 u}{\partial x^2}$$

using operator/time splitting. For the advection term use a forward-in-time, backward-in-space (upwind) scheme with a time interval of 30 s, while for the diffusion term use the forward-in-time scheme with a time step of 120 s. Use a grid with 601 points and a grid spacing of 1 km, an advection speed of 15 m s^{-1}, and a diffusion coefficient of 4000 m^2 s^{-1}. Initialize the solution with a box of width 20 grid points, centered on the grid. After 250 minutes (500 advection iterations and 125 diffusion iterations) your results should look like the figure below. Note that if the diffusion coefficient is set to zero and the model rerun, the results will look like those of Ex. 5.1.

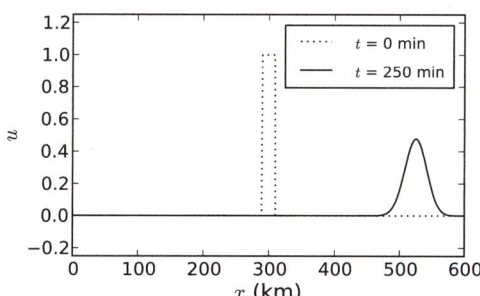

CHAPTER 6

The Filtered Equations I: Concepts

6.1   Purpose of Filtered Equation Models

Lewis Fry Richardson was the first to apply finite-difference methods to the primitive equations in an attempt to predict the future state of the atmosphere. His attempt failed, primarily due to imbalances in the initial conditions that he used.

On the synoptic scale in midlatitudes the atmosphere is nearly in gradient and hydrostatic balance. Processes such as latent heat release or radiative heating and cooling disturb these balances, resulting in the generation of gravity waves which propagate away from the disturbance and restore balance by redistributing mass and momentum. This process is referred to as geostrophic adjustment.[1]

Richardson obtained his initial conditions from actual observations, and presumably the atmosphere at the time was in balance;

---

[1]Geostrophic balance is achieved when the horizontal pressure gradient acceleration exactly balances the Coriolis acceleration, and is appropriate for steady, straight-line, frictionless flow. In steady, curved flow the two accelerations cannot be in balance, but must differ by an amount equal to the centripetal acceleration. This is known as gradient balance. Because atmospheric flows are often curved and so are closer to gradient balance than geostrophic balance, the term gradient adjustment is actually more appropriate. However, the term geostrophic adjustment is more commonly used.

however, there are always errors present in observational data. The observations must also be interpolated onto the model grid, which introduces further error. Though the total error may be small, even a slight imbalance can generate unrealistically large-amplitude gravity waves, swamping any meteorologically relevant pressure tendency. These large-amplitude gravity waves doomed Richardson's forecasting attempt.[2]

The problem of balancing the initial conditions was not solved until the 1960s. Before this, a different approach to numerical weather prediction had to be employed. Instead of the full set of governing equations, a reduced set of equations based on quasigeostrophic (QG) theory was used. The quasigeostrophic equations do not support gravity wave propagation, and therefore the initial conditions do not have to be balanced. Models based on quasigeostrophic theory are called filtered-equation models because gravity waves and acoustic waves are both filtered out, leaving only Rossby waves as solutions.[3]

The remainder of this chapter is devoted to explaining the concepts of the quasigeostrophic vorticity equation, which is the predictive equation used in filtered-equation models. In Chapter 7 we will learn how to solve this equation numerically.

6.2   The Quasigeostrophic Vorticity Equation

The actual wind is a sum of the geostrophic wind and the ageostrophic wind,

$$\vec{V} = \vec{V}_g + \vec{V}_a. \tag{6.1}$$

The ageostrophic wind is usually of much smaller magnitude than the geostrophic wind, and so on the large scale the geostrophic wind can often be used as a proxy for the actual wind. Quasigeostrophic

---

[2]Even if Richardson had been able to initialize his simulation with balanced conditions, his simulation would have eventually failed because the time increment he used was too large, violating the Courant-Friedrich-Lewy stability condition.

[3]Even after the advent of viable primitive equation models in the mid-1960s, filtered-equation models continued to be run operationally at the U. S. National Centers for Environmental Prediction (NCEP) throughout the 1970s. Barotropic QG models were used to extend the 500 hPa forecasts from the PE models, and a stand-alone barotropic model continued to be run until 1983.

theory capitalizes upon this principle by replacing the actual wind with the geostrophic wind in the governing equations, except in terms that are used for computing divergence. The reason for using the actual wind in the divergence term is that the geostrophic wind is nondivergent. If the geostrophic wind were used in the divergence term, then no mechanism would exist in the equations for generating vertical motion or spin-up of absolute vorticity.

The quasigeostrophic vorticity equation is

$$\frac{\partial \zeta_g}{\partial t} + \vec{V}_g \cdot \nabla \zeta_g + \beta v_g = -f \nabla \cdot \vec{V}_a, \tag{6.2}$$

where the subscripts g and a refer to the geostrophic and ageostrophic winds. The terms in this equation represent the local tendency of relative vorticity, advection of relative vorticity, advection of planetary vorticity, and divergence. In quasigeostrophic theory there are only three means by which the vorticity at a fixed point in space may change: 1) advection of relative vorticity by the geostrophic wind; 2) advection of planetary vorticity by the geostrophic wind; and 3) divergence of the ageostrophic wind.

6.3   Alternate Forms of the QG Vorticity Equation

The QG vorticity equation can be written in several completely equivalent forms. Using the continuity equation in pressure coordinates,

$$\nabla \cdot \vec{V}_a + \frac{\partial \omega}{\partial p} = 0, \tag{6.3}$$

the right-hand side of (6.2) is altered to

$$\frac{\partial \zeta_g}{\partial t} + \vec{V}_g \cdot \nabla \zeta_g + \beta v_g = f \frac{\partial \omega}{\partial p}. \tag{6.4}$$

Defining the geostrophic total derivative as

$$\frac{D_g}{Dt} \equiv \frac{\partial}{\partial t} + \vec{V}_g \cdot \nabla = \frac{\partial}{\partial t} + u_g \frac{\partial}{\partial x} + v_g \frac{\partial}{\partial y}, \tag{6.5}$$

we can write (6.4) as

$$\frac{D_g \zeta_g}{Dt} = -\beta v_g + f \frac{\partial \omega}{\partial p}. \tag{6.6}$$

In this form we see that in quasigeostrophic theory the relative vorticity following a fluid parcel can only change via advection of planetary vorticity or through divergence. Yet another, completely equivalent form is

$$\frac{D_g \eta_g}{Dt} = f\frac{\partial \omega}{\partial p}, \tag{6.7}$$

where η_g is the geostrophic absolute vorticity ($\zeta_g + f$). Equation (6.7) is interpreted as meaning that the only mechanism for changing the absolute vorticity of a fluid parcel is divergence or convergence.

6.4   The Barotropic QG Vorticity Equation

In a barotropic fluid the pressure and density surfaces are everywhere parallel. Thus, there is no thermal wind and so no vertical wind shear. The velocity is therefore the same at all levels. We form the barotropic QG vorticity equation by integrating (6.4) from the top of the atmosphere to the surface,

$$\int_0^{p_s} \left(\frac{\partial \zeta_g}{\partial t} + \vec{V}_g \cdot \nabla \zeta_g + \beta v_g \right) dp = \int_0^{p_s} f\frac{\partial \omega}{\partial p} dp, \tag{6.8}$$

resulting in (see Exercise 6.1)

$$\frac{\partial \zeta_g}{\partial t} + \vec{V}_g \cdot \nabla \zeta_g + \beta v_g = \frac{f\omega_s}{p_s}. \tag{6.9}$$

In this equation ω_s is the pressure-coordinate vertical velocity at the surface and p_s is the surface pressure. Equation (6.9) is valid at any vertical level within the fluid, since the velocity and vorticity are not dependent on height.

The right-hand side of (6.9) represents divergence, but in a barotropic QG fluid the divergence is due solely to the vertical velocity forced by the flow along the surface. In flat terrain ω_s would be zero, and so would the divergence term. The vertical velocity at the surface is expressed as (see Exercise 6.2)

$$\omega_s = -\rho_s g \vec{V}_g \cdot \nabla z_s, \tag{6.10}$$

where ρ_s is the density at the surface. Using (6.10) in (6.9), along with the ideal gas law, results in yet another form of the barotropic

QG vorticity equation,

$$\frac{\partial \zeta_g}{\partial t} + \vec{V}_g \cdot \nabla \zeta_g + \beta v_g = -fgR_dT_s\vec{V}_g \cdot \nabla z_s, \qquad (6.11)$$

where T_s is the surface temperature. Note the following two very important points about the barotropic QG vorticity equation:

- Relative vorticity following a fluid parcel can only change in response to changes in latitude or due to sloped terrain;

- The absolute vorticity following a fluid parcel can only change in response to sloped terrain.

Over flat terrain absolute vorticity is conserved following a parcel. This implies that the only type of cyclogenesis that a barotropic QG model is capable of representing is that due to terrain, such as lee-side troughs. Development of cyclones due to baroclinic instability cannot occur in a barotropic QG model.

6.5   The Equivalent-barotropic Vorticity Equation

An *equivalent-barotropic* fluid is one in which the density and pressure surfaces intersect, but only such that the thermal wind is always directed along the height contours. This means that the wind speed can vary with height, but the direction cannot change with height except by exactly 180 degrees. To apply the concept of equivalent barotropy to the quasigeostrophic vorticity equation, we first define the vertically-averaged wind vector,

$$\overline{\vec{V}}_g = \frac{1}{p_s} \int_0^{p_s} \vec{V}_g dp. \qquad (6.12)$$

We then assume that the geostrophic wind vector at any pressure level p can be written in terms of the vertically-average wind vector multiplied by a pressure dependent multiplier, $A(p)$, such that

$$\vec{V}_g(x,y,p,t) = A\overline{\vec{V}}_g(x,y,t). \qquad (6.13)$$

Using (6.13) in (6.4) gives

$$A\frac{\partial \overline{\zeta}_g}{\partial t} + A^2\overline{\vec{V}}_g \cdot \nabla \overline{\zeta}_g + A\beta \overline{v}_g = f\frac{\partial \omega}{\partial p}, \qquad (6.14)$$

where the over-bar symbol indicates a vertical average. Integrating (6.14) throughout the depth of the atmosphere gives

$$\frac{\partial \overline{\zeta_g}}{\partial t} \int_0^{p_s} A\, dp + \overline{\vec{V}_g} \cdot \nabla \overline{\zeta_g} \int_0^{p_s} A^2\, dp$$

$$+ \beta \overline{v}_g \int_0^{p_s} A\, dp = f \int_0^{p_s} \frac{\partial \omega}{\partial p}\, dp \quad (6.15)$$

(remember that all over-barred quantities are independent of pressure and can be moved outside of the integral). The multiplier $A(p)$ is normalized such that

$$\frac{1}{p_s} \int_0^{p_s} A\, dp = 1, \quad (6.16)$$

so that (6.15) simplifies to

$$\frac{\partial \overline{\zeta_g}}{\partial t} + \overline{A^2}\, \overline{\vec{V}_g} \cdot \nabla \overline{\zeta_g} + \beta \overline{v}_g = \frac{f \omega_s}{p_s}. \quad (6.17)$$

Multiplying through by $\overline{A^2}$ gives

$$\overline{A^2} \frac{\partial \overline{\zeta_g}}{\partial t} + \overline{A^2}\, \overline{\vec{V}_g} \cdot \overline{A^2}\, \nabla \overline{\zeta_g} + \overline{A^2}\, \beta \overline{v}_g = \overline{A^2} \frac{f \omega_s}{p_s}. \quad (6.18)$$

There must exist some pressure level p^* at which the geostrophic wind is given by

$$\vec{V}_g(x, y, p^*, t) = \overline{A^2}\, \overline{\vec{V}_g}(x, y, t), \quad (6.19)$$

and at this level (6.18) becomes

$$\frac{\partial \zeta_g}{\partial t} + \vec{V}_g \cdot \nabla \zeta_g + \beta v_g = \overline{A^2} \frac{f \omega_s}{p_s}. \quad (6.20)$$

Equation (6.20) is called the *equivalent-barotropic QG vorticity equation*. It looks very much like the barotropic QG vorticity equation, (6.9), with the exception of the $\overline{A^2}$ on the right-hand side. Also, notice that if the terrain is flat ($\omega_s = 0$), the equivalent-barotropic QG vorticity equation and barotropic QG vorticity equation are identical.

A form of (6.20) was the basis for the first successful operational numerical weather prediction model. Models based on this equation are called equivalent-barotropic QG models, or often simply barotropic QG models. When applying the equivalent-barotropic QG vorticity equation to the atmosphere it is important to know at which level (6.20) is valid. In most instances $\overline{A^2}$ has a value of approximately 1.25, and so p^* is approximately 600 hPa. A common assumption is that the level-of-nondivergence is around 600 hPa. This is no coincidence; the equivalent-barotropic QG vorticity equation allows divergence only through the vertical motion induced by flow over terrain, and so would be expected to be most accurate at the level of nondivergence. Although the level of nondivergence is closer to the 600 hPa level, 500 hPa is traditionally taken as its proxy. Thus, barotropic QG models are normally assumed to represent conditions at 500 hPa.

As with the barotropic QG vorticity equation, the equivalent-barotropic QG vorticity equation cannot develop new circulations except in response to flow over terrain. Thus, it can simulate the development of a lee-side trough, but it cannot develop new cyclones through baroclinic instability.

6.6   Writing Advection in Jacobian Form

The relative vorticity advection term in the QG vorticity equation has the form

$$\vec{V}_g \cdot \nabla \zeta_g = u_g \frac{\partial \zeta_g}{\partial x} + v_g \frac{\partial \zeta_g}{\partial y}. \tag{6.21}$$

The geostrophic wind can be written in terms of the streamfunction, ψ, as

$$u_g = -\frac{\partial \psi}{\partial y}; \quad v_g = \frac{\partial \psi}{\partial x}, \tag{6.22}$$

so that the relative vorticity advection term becomes

$$\vec{V}_g \cdot \nabla \zeta_g = \frac{\partial \psi}{\partial x} \frac{\partial \zeta_g}{\partial y} - \frac{\partial \psi}{\partial y} \frac{\partial \zeta_g}{\partial x}. \tag{6.23}$$

This has the form of the Jacobian operator, defined as

$$J(a, b) = \frac{\partial a}{\partial x} \frac{\partial b}{\partial y} - \frac{\partial a}{\partial y} \frac{\partial b}{\partial x}. \tag{6.24}$$

Vorticity advection therefore can be written as

$$\vec{V}_g \cdot \nabla \zeta_g = J\left(\psi, \zeta_g\right). \tag{6.25}$$

In a similar manner we can write

$$\vec{V}_g \cdot \nabla z_s = J\left(\psi, z_s\right), \tag{6.26}$$

so that the barotropic QG vorticity equation can be written as

$$\frac{\partial}{\partial t}\zeta_g + J\left(\psi, \zeta_g\right) + \beta v_g = -fgR_dT_sJ\left(\psi, z_s\right). \tag{6.27}$$

And finally, the geostrophic vorticity can be written in terms of streamfunction,

$$\zeta_g = \nabla^2\psi, \tag{6.28}$$

which allows the barotropic QG vorticity equation to be written with streamfunction as the sole dependent variable,

$$\frac{\partial}{\partial t}\nabla^2\psi + J\left(\psi, \nabla^2\psi\right) + \beta\frac{\partial\psi}{\partial x} = -fgR_dT_sJ\left(\psi, z_s\right). \tag{6.29}$$

6.7   Enstrophy and Other Domain-invariant Quantities

The barotropic QG vorticity equation with no topography can be written in flux-form as

$$\frac{\partial \eta_g}{\partial t} = -\nabla \cdot \left(\eta_g \vec{V}_g\right). \tag{6.30}$$

Integrating (6.30) over a 2D domain results in

$$\frac{\partial}{\partial t}\iint \eta_g dx dy = -\iint \nabla \cdot \left(\eta_g \vec{V}_g\right) dx dy. \tag{6.31}$$

Gauss' Divergence Theorem, expressed mathematically as

$$\iint \nabla \cdot \vec{B}\, dx dy = \oint \vec{B} \cdot \hat{n}\, ds \tag{6.32}$$

where $\vec{B}$ is any vector and $\hat{n}$ is a unit vector perpendicular to the boundary of the domain, states that the integral of a divergence over a closed domain is equal to the integral of the flux around the perimeter of the domain. Applying this theorem to (6.31) yields

$$\frac{\partial}{\partial t} \iint \eta_g \, dxdy = - \oint \left(\eta_g \vec{V} \right) \cdot \hat{n} \, ds.$$
(6.33)

The domain-averaged vorticity is

$$\bar{\eta}_g = \iint \eta_g \, dxdy \Big/ \text{Area}$$
(6.34)

and so the left-hand side of (6.33) is proportional to the tendency of the domain-averaged vorticity. The right-hand side is the total flux of vorticity across the domain boundary. If the domain is closed, there are no fluxes across the boundary and this term is zero. The result is then

$$\frac{\partial}{\partial t} \bar{\eta}_g = \frac{\partial}{\partial t} \left(\bar{\zeta}_g + \bar{f} \right) = \frac{\partial}{\partial t} \bar{\zeta}_g = 0$$
(6.35)

(recall that f is not a function of t). This equation shows that the mean vorticity over the domain remains constant and is therefore a *domain-invariant property*.

Two other important domain-invariant properties are mean enstrophy, defined as $\frac{1}{2}\overline{\zeta_g^2}$, and mean kinetic energy (see Exercises 6.5 and 6.6). In a barotropic QG atmosphere the means of vorticity, enstrophy, and kinetic energy remain constant. When building a model based on barotropic QG vorticity equation, care must be taken to design the model such that these domain-invariant properties are conserved.

6.8 Baroclinic Filtered-equation Models

Even though the first operational numerical weather prediction model was based on the barotropic QG vorticity equation and was successful for many years in forecasting conditions at 500 hPa, it suffered from not being able to represent cyclogenesis due to baroclinic instability. This led to the development of multiple-level filtered-equation models. Such models are based on both the QG vorticity equation

$$\frac{\partial}{\partial t} \nabla^2 \psi + J(\psi, \nabla^2 \psi) + \beta \frac{\partial \psi}{\partial x} = f_0 \frac{\partial \omega}{\partial p}$$
(6.36)

and the QG thermodynamic energy equation (see Exercise 6.7),

$$\frac{\partial}{\partial t}\left(\frac{\partial \psi}{\partial p}\right) + J\left(\psi, \nabla \frac{\partial \psi}{\partial p}\right) + \frac{\sigma}{f_0}\omega = -\frac{R_d}{c_p f_0 p}\dot{Q}. \qquad (6.37)$$

The terms on the left-hand side of (6.37), in order from left to right, represent 1) the local temperature tendency, 2) horizontal thermal advection, and 3) combined vertical thermal advection and adiabatic heating and cooling. The term on the right-hand side represents the diabatic heating and cooling rate and can include processes such as radiation and latent and sensible heating and cooling. The parameter σ is the stability parameter given by

$$\sigma = -\frac{1}{\rho}\frac{\partial \ln \theta}{\partial p}, \qquad (6.38)$$

and is negative for statically stable conditions.

Equations (6.36) and (6.37) form a closed set of equations with ψ and ω as the dependent variables, and so can be numerically integrated to predict the future state of the streamfunction field. These equations are applied at multiple vertical levels to form a QG model capable of representing baroclinic instability. This application is briefly illustrated in Chapter 7.

Exercises

Ex. 6.1:   Perform the integration in (6.8) to obtain (6.9).

Ex. 6.2:   Show that $\omega_s = -\rho_s g \vec{V}_g \cdot \nabla z_s$.

Ex. 6.3:   Show that $\vec{V}_g \cdot \nabla z_s = J(\psi, z_s)$.

Ex. 6.4:   Show that the barotropic QG vorticity equation with no terrain can be written as

$$\frac{\partial \eta_g}{\partial t} = -\nabla \cdot \left(\eta_g \vec{V}_g\right). \qquad (6.39)$$

Hint: Recall that the geostrophic wind is nondivergent.

   Exercises 6.5 and 6.6 guide the reader through derivations of the domain invariants of mean enstrophy and mean kinetic energy.

Ex. 6.5:   Show that mean enstrophy is a domain invariant through the following steps:

a. Multiply both sides of (6.30) by η_g and integrate over the domain to get

$$\iint \frac{\partial}{\partial t}\left(\frac{\eta_g^2}{2}\right) dxdy = -\iint \nabla \cdot \left(\frac{\eta_g^2}{2}\vec{V}_g\right) dxdy. \qquad (6.40)$$

b. Apply Gauss' divergence theorem to the right-hand side of (6.40) to get

$$\frac{\partial}{\partial t}\iint\left(\frac{\eta_g^2}{2}\right) dxdy = 0. \qquad (6.41)$$

c. Expand the absolute vorticity in (6.41) to yield

$$\frac{\partial}{\partial t}\iint \frac{\zeta_g^2}{2}dxdy + \frac{\partial}{\partial t}\iint f\zeta_g dxdy + \frac{\partial}{\partial t}\iint \frac{f^2}{2}dxdy = 0. \qquad (6.42)$$

d. Show that the second and third terms are zero, so that we are left with

$$\frac{\overline{\zeta_g^2}}{2} = 0. \qquad (6.43)$$

Ex. 6.6: Show that mean kinetic energy is a domain invariant through the following steps:

a. Multiply both sides of (6.30) by the streamfunction and then integrate over the domain to get

$$\iint \psi\frac{\partial \eta_g}{\partial t}dxdy = -\iint \psi\nabla \cdot \left(\eta_g\vec{V}_g\right) dxdy. \qquad (6.44)$$

b. Show that

$$\psi\frac{\partial \eta_g}{\partial t} = \nabla \cdot \left(\psi\nabla\frac{\partial\psi}{\partial t}\right) - \frac{\partial}{\partial t}\left(\frac{\nabla\psi \cdot \nabla\psi}{2}\right). \qquad (6.45)$$

c. Show that

$$\psi\nabla \cdot \left(\eta_g \vec{V}_g\right) = \nabla \cdot \left(\psi\eta_g \vec{V}_g\right) - \eta_g\vec{V}_g \cdot \nabla\psi. \qquad (6.46)$$

d. Using (6.45) and (6.46) show that (6.44) can be written as

$$\frac{\partial}{\partial t}\iint \left(\frac{\nabla\psi \cdot \nabla\psi}{2}\right) dxdy = \iint \nabla \cdot \left(\psi\nabla\frac{\partial\psi}{\partial t}\right) dxdy$$
$$+ \iint \nabla \cdot \left(\psi\eta_g\vec{V}_g\right) dxdy - \iint \eta_g\vec{V}_g \cdot \nabla\psi dxdy. \qquad (6.47)$$

e. Explain why the last term in (6.47) is zero. Hint: Think of the relationship between geostrophic wind and streamfunction.

f. Use Gauss' divergence theorem to explain why the remaining terms on the right-hand side of (6.47) are zero.

g. Explain how $\frac{1}{2}\nabla\psi \cdot \nabla\psi$ is kinetic energy per unit mass (*KE*), and therefore, we have shown that

$$\frac{\partial}{\partial t}\overline{KE} = 0. \tag{6.48}$$

Ex. 6.7:   The QG thermodynamic energy equation is

$$\frac{\partial T}{\partial t} + J(\psi, T) - \frac{\sigma p}{R_d}\omega = \frac{\dot{Q}}{c_p}. \tag{6.49}$$

Using the hydrostatic equation in pressure coordinates and the ideal gas law, show that temperature can be written as

$$T = -\frac{p f_0}{R_d}\frac{\partial \psi}{\partial p} \tag{6.50}$$

and that therefore the thermodynamic energy equation can be written in terms of streamfunction as

$$\frac{\partial}{\partial t}\left(\frac{\partial \psi}{\partial p}\right) + J\left(\psi, \nabla\frac{\partial \psi}{\partial p}\right) + \frac{\sigma}{f_0}\omega = -\frac{R_d}{c_p f_0 p}\dot{Q}. \tag{6.51}$$

## CHAPTER 7

The Filtered Equations II: Numerical Methods

7.1 The QG Barotropic Model

In this chapter we describe a method for numerically solving the barotropic or equivalent-barotropic QG vorticity equation. For simplicity we assume a flat topography so that the equation to solve is

$$\frac{\partial \zeta_g}{\partial t} + J\left(\psi, \zeta_g\right) + \beta v_g = 0. \tag{7.1}$$

The conceptual method for the QG barotropic model begins with an initial streamfunction field, which is used to determine the initial vorticity field. Equation (7.1) is then numerically integrated forward-in-time in order to predict the future vorticity, and hence, the future streamfunction.[1] The steps for solution are:

Step 1: From the initial streamfunction, calculate the initial geostrophic vorticity using the relationship

$$\zeta_g = \nabla^2 \psi. \tag{7.2}$$

---

[1]The streamfunction is directly proportional to geopotential height, so knowledge of the streamfunction field is tantamount to knowledge of the geopotential height field.

Step 2: Calculate the relative vorticity advection, $J(\psi, \zeta)$. Also, calculate the initial meridional geostrophic wind using

$$v_g = \frac{\partial \psi}{\partial x}. \tag{7.3}$$

Step 3: Use finite differencing to solve (7.1) for the future value of the geostrophic vorticity, ζ_g. Using the leapfrog time-differencing scheme the recursion relation is

$$\zeta_{i,j}^{n+1} = \zeta_{i,j}^{n-1} - 2\Delta t \left([J(\psi, \zeta)]_{i,j}^n + \beta v_{i,j}^n \right), \tag{7.4}$$

where the subscripts i and j refer to the x and y grid-point indices, and n refers to the time step index.

Step 4: Use (7.2) to find the new streamfunction from the new vorticity value.

Steps 2 through 4 are then repeated iteratively to push the predicted values further and further into the future.

Although the method of solution is conceptually simple, there are some complications that arise. These are:

- Finding the correct finite-difference form for the Laplacian in order to find the vorticity from the streamfunction using (7.2).

- Finding the correct finite-difference form of the Jacobian so that the domain invariants (mean vorticity, enstrophy, and kinetic energy) are conserved.

- Inverting (7.2) so that once the new value of vorticity is found the new value of the streamfunction can be obtained.

In the following sections we address these three issues as well as other considerations for filtered-equation models.

7.2 Finite-difference Form of the Laplacian

The two-dimensional Laplacian in Cartesian coordinates is defined as

$$\nabla^2 = \frac{\partial^2}{\partial x^2} + \frac{\partial^2}{\partial y^2}, \tag{7.5}$$

and can be approximated by adding the finite-difference forms of the second derivatives for both the x and y directions. For a uniform grid of interval d in both the x and y directions, this approximation is

$$\nabla^2 \psi_{i,j} = \frac{\psi_{i+1,j} + \psi_{i-1,j} + \psi_{i,j+1} + \psi_{i,j-1} - 4\psi_{i,j}}{d^2}, \tag{7.6}$$

and is known as the *five-point Laplacian stencil* because it uses information at five grid points. It is second-order accurate. A slightly more accurate, but still second-order, approximation uses information at nine grid points, and is given by

$$\nabla^2 \psi_{i,j} = \frac{1}{6d^2} \begin{bmatrix} 4\left(\psi_{i+1,j} + \psi_{i-1,j} + \psi_{i,j+1} + \psi_{i,j-1}\right) \\ + \psi_{i+1,j+1} + \psi_{i+1,j-1} + \psi_{i-1,j-1} \\ + \psi_{i-1,j+1} - 20\psi_{i,j} \end{bmatrix}. \tag{7.7}$$

Note that neither (7.6) nor (7.7) can be used to directly calculate the Laplacian on the boundary of the grid. On the boundaries the vorticity must be assumed to be zero, or if cyclic boundary conditions (discussed later in this chapter) are used a modified form of (7.6) or (7.7) can be applied.

7.3 Finite-difference Form of the Jacobian

Three completely equivalent ways of writing the Jacobian are:

$$J(\psi, \zeta) = \frac{\partial \psi}{\partial x} \frac{\partial \zeta}{\partial y} - \frac{\partial \psi}{\partial y} \frac{\partial \zeta}{\partial x} \tag{7.8}$$

$$J(\psi, \zeta) = \frac{\partial}{\partial x}\left(\psi \frac{\partial \zeta}{\partial y}\right) - \frac{\partial}{\partial y}\left(\psi \frac{\partial \zeta}{\partial x}\right) \tag{7.9}$$

$$J(\psi, \zeta) = \frac{\partial}{\partial y}\left(\zeta \frac{\partial \psi}{\partial x}\right) - \frac{\partial}{\partial x}\left(\zeta \frac{\partial \psi}{\partial y}\right). \tag{7.10}$$

Corresponding to these are three ways of expressing the Jacobian in finite-difference form:

$$J_{i,j}^{I} = \frac{1}{4d^2} \begin{bmatrix} (\psi_{i+1,j} - \psi_{i-1,j})(\zeta_{i,j+1} - \zeta_{i,j-1}) \\ - (\psi_{i,j+1} - \psi_{i,j-1})(\zeta_{i+1,j} - \zeta_{i-1,j}) \end{bmatrix} \tag{7.11}$$

$$J_{i,j}^{II} = \frac{1}{4d^2} \begin{bmatrix} \psi_{i+1,j}(\zeta_{i+1,j+1} - \zeta_{i+1,j-1}) \\ - \psi_{i-1,j}(\zeta_{i-1,j+1} - \zeta_{i-1,j-1}) \\ - \psi_{i,j+1}(\zeta_{i+1,j+1} - \zeta_{i-1,j+1}) \\ + \psi_{i,j-1}(\zeta_{i+1,j-1} - \zeta_{i-1,j-1}) \end{bmatrix} \tag{7.12}$$

$$J_{i,j}^{III} = \frac{1}{4d^2} \begin{bmatrix} \zeta_{i,j+1}(\psi_{i+1,j+1} - \psi_{i-1,j+1}) \\ - \zeta_{i,j-1}(\psi_{i+1,j-1} - \psi_{i-1,j-1}) \\ - \zeta_{i+1,j}(\psi_{i+1,j+1} - \psi_{i+1,j-1}) \\ + \zeta_{i-1,j}(\psi_{i-1,j+1} - \psi_{i-1,j-1}) \end{bmatrix}. \tag{7.13}$$

None of these finite-difference approximations for the Jacobian conserves both kinetic energy and enstrophy over the model domain, and they are therefore unsuitable for use by themselves in the model. However, Arakawa showed that by averaging the results from all three, the domain invariants are conserved. Our finite-difference approximation for the Jacobian will therefore be

$$J_{i,j} = \frac{J_{i,j}^{I} + J_{i,j}^{II} + J_{i,j}^{III}}{3}. \tag{7.14}$$

Unless cyclic boundary conditions are used, the approximations (7.11), (7.12), and (7.13) cannot be applied on the boundary of the grid, nor on any grid point adjacent to the boundary. This is because, when calculating the Jacobian at a certain grid point, the vorticities at adjacent grid points are needed and, as previously discussed, the vorticity cannot be calculated on the boundary of the grid unless cyclic boundary conditions are used.

7.4   Solving Poisson's Equation

Now that we know how to represent the Jacobian as a finite difference our last hurdle to creating a barotropic model is to figure out how to invert (7.2) to get the new streamfunction at time level $n+1$ from the vorticity we calculated for time level $n+1$. In essence, we have to take an 'inverse Laplacian.' Equation (7.2), rewritten as

$$\nabla^2 \psi = \zeta, \tag{7.15}$$

is an elliptic equation known as *Poisson's equation*. Note that in this instance the right-hand side of the equation (the vorticity) is a known quantity, and we want to find the streamfunction such that the Laplacian of the streamfunction will yield the known vorticity value.

Elliptic equations are notoriously time-consuming to solve numerically. We will discuss two methods for solving (7.15). These are the *method of relaxation* and the *matrix method*. Though our discussion will be confined to Poisson's equation, the concepts can be applied to any elliptic partial differential equation.

7.4.1 Method of relaxation

The finite-difference form of (7.15), using the five-point stencil for the Laplacian, is

$$\frac{\psi_{i-1,j} + \psi_{i+1,j} + \psi_{i,j-1} + \psi_{i,j+1} - 4\psi_{i,j}}{d^2} - \zeta_{i,j} = 0. \qquad (7.16)$$

We have not explicitly written the time-level index since all quantities are taken at the same time level. We solve (7.16) by first making a guess as to what the streamfunction should be at each grid point. Our guess is denoted as G. We substitute this guess in place of the streamfunction in (7.16), and hope that in doing so the right-hand side remains zero for every grid point. If this is true, then we guessed the correct solution and we are done! However, it is unlikely that our guess will be correct, and so instead the right-hand side will be a nonzero number called the *residual*. The residual at each grid point is the difference between the Laplacian of our guess field at the grid point and the vorticity at the grid point, and is expressed as

$$\frac{G_{i-1,j} + G_{i+1,j} + G_{i,j-1} + G_{i,j+1} - 4G_{i,j}}{d^2} - \zeta_{i,j} = R_{i,j}. \qquad (7.17)$$

By adding a value, denoted as γ, to our guess at grid point (i,j) we can make the residual vanish at that point,

$$\frac{G_{i-1,j} + G_{i+1,j} + G_{i,j-1} + G_{i,j+1} - 4\left(G_{i,j} + \gamma\right)}{d^2} - \zeta_{i,j} = 0. \qquad (7.18)$$

The value of γ required for eliminating the residual is found by solving (7.18) for γ and also using the definition of the residual from

(7.17). The result is

$$\gamma = \frac{d^2}{4} R_{i,j}.$$ (7.19)

So, to make the residual disappear at any grid point (i, j) we replace the guess at that grid point using

$$G_{i,j} \leftarrow G_{i,j} + \frac{d^2}{4} R_{i,j}.$$ (7.20)

Notice, however that the residual at grid point (i, j) depends on the value of the guess at adjacent grid points. So having eliminated the residual at grid point (i, j), if we then try to do the same thing at the next grid point $(i + 1, j)$ the residual at (i, j) will become nonzero again and we will have just undone what we previously had accomplished. The good news is that even though the residual at (i, j) is no longer zero, it is likely to be smaller than what it was for the previous guess. In the method of relaxation we continually iterate through the grid, updating the guesses at each grid point using (7.20), and keep doing this as many times as necessary until no grid point has a residual with magnitude greater than some threshold value. Once this is achieved, the final guess field will be close to the solution of the streamfunction for (7.15).

The general algorithm for relaxation is diagrammed in Fig. 7.1, and the steps are summarized as follows:

1. Assume a first guess field for the streamfunction, G^0, at every grid point. The subscript '0' indicates that this is the initial guess.

2. Iterate through the interior grid points of the grid calculating the initial residual at each grid point using

$$R_{i,j}^0 = \nabla^2 G_{i,j}^0 - \zeta_{i,j}.$$ (7.21)

3. Check if any residual in the grid has a magnitude greater than the threshold value. If there are no residuals exceeding the threshold, then the guess field G^0 is the streamfunction that solves (7.15). If there are still large residuals, then proceed with Step 4 below.

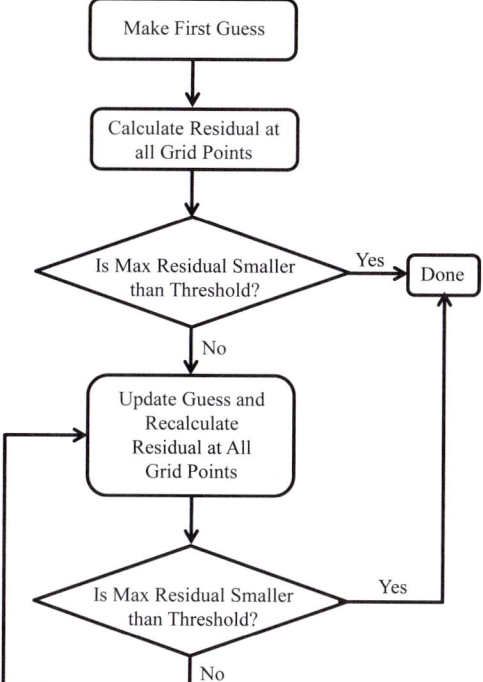

Fig. 7.1: Flowchart showing general algorithm for the method of relaxation.

4. Iterate through the grid again, this time replacing the value at each grid point with a new value determined by

$$G_{i,j}^1 = G_{i,j}^0 + \frac{d^2}{4} R_{i,j}^0. \qquad (7.22)$$

Before moving to the next grid point determine the new residual by calculating the new Laplacian and subtracting the vorticity,

$$R_{i,j}^1 = \nabla^2 G_{i,j}^1 - \zeta_{i,j}. \qquad (7.23)$$

5. After Step 4 has been performed at every grid point check again to see if any residual R^1 has a magnitude greater than the threshold. If not, the latest guess G^1 is the streamfunction solution to (7.15) and we stop. If the residuals are still too large, then continue to step 6 below.

6. Iterate again through the grid replacing the old value of the guess (superscript m) at each grid point with a new value (superscript $m + 1$) determined by

$$G_{i,j}^{m+1} = G_{i,j}^m + \frac{d^2}{4} R_{i,j}^m, \tag{7.24}$$

and before moving to the next grid point determine the new residual by calculating the Laplacian and subtracting the vorticity,

$$R_{i,j}^{m+1} = \nabla^2 G_{i,j}^{m+1} - \zeta_{i,j}. \tag{7.25}$$

7. Check to see if any residual R^{m+1} has a magnitude greater than the threshold. If not, the latest guess G^{m+1} is the streamfunction solution to (7.15). If the residuals are still too large, then keep repeating Steps 6 and 7 until the residuals have small enough magnitudes.

Further comments on the method of relaxation:

- When iterating through the grid, only iterate through, and change values on, the interior grid points unless cyclic boundary conditions are used. Do not change any values on fixed boundaries.[2]

- When calculating the Laplacian at a grid point the values of the field at adjacent grid points may be either the old guess or an updated guess, because some of the adjacent grid points will have been updated while others will not have been updated. You will be using a mixture of old guesses and new guesses when calculating the Laplacian. Mixing the guess fields like this may seem counterintuitive, but it is correct.

- Since it does not matter what initial guess field you choose in the interior of the domain you may as well choose something that might be close to the actual solution (perhaps a previous

---

[2] This is because the solution to Poisson's equation is completely and uniquely determined by the boundary conditions. The boundary values must remain fixed during the relaxation process. The streamfunction that is picked for the boundary condition will uniquely determine the streamfunction in the interior of the domain.

analysis of the streamfunction). Choosing a reasonable guess field can reduce the number of iterations needed to achieve a solution.

- The solution will converge faster (residuals decrease in magnitude more rapidly) if instead of using (7.24) for the updated guess, we instead use

$$G_{i,j}^{m+1} = G_{i,j}^m + \alpha \frac{d^2}{4} R_{i,j}^m. \qquad (7.26)$$

In this equation the coefficient α is determined by

$$\alpha = \frac{8 - 4\sqrt{4 - t^2}}{t^2}$$
$$t = \cos\left(\frac{\pi}{M}\right) + \cos\left(\frac{\pi}{N}\right) \qquad (7.27)$$

where M and N are the number of grid points in the x and y directions respectively. This method is called *over-relaxation*, and the parameter α is called the *over-relaxation coefficient*.

7.4.2   Matrix and spectral methods

Another possibility for solving Poisson's equation numerically is to explicitly write (7.16) for every grid point. The total number of grid points on a two-dimensional grid is $M = NX \times NY$, where NX and NY are the number of grid points in the x and y directions. We would then have a total of M simultaneous equations with M unknowns. The unknowns are the streamfunction values at each grid point. This system of equations can be written as a matrix equation and solved using existing numerical algorithms. Another choice for solving elliptic equations numerically is a method involving Fourier transforms (*spectral methods*). Details of these methods are beyond the scope of this book.

7.5   General Algorithm for the Barotropic QG Model

In this section we give a more detailed algorithm for programming the barotropic QG model. The general algorithm is laid out in Fig. 7.2. The specific steps are:

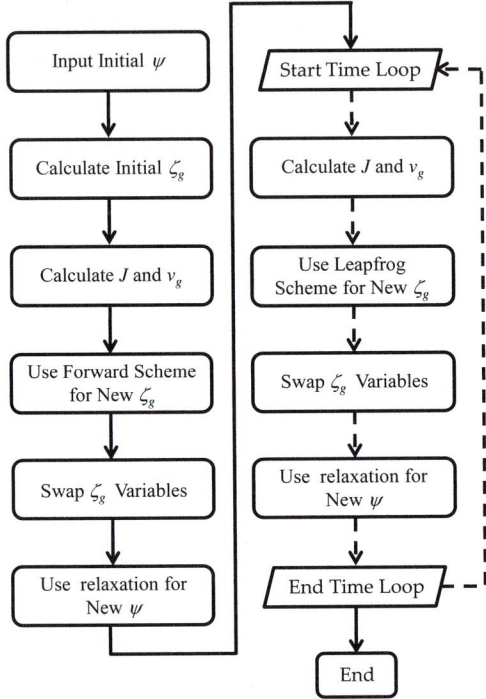

Fig. 7.2: Flowchart showing general algorithm for solving the barotropic model.

Step 1: Input the initial streamfunction, ψ^0.

Step 2: Calculate the initial vorticity, ζ^0, using (7.2).

Step 3: Calculate the Jacobian using (7.14), and the initial meridional geostrophic wind using (7.3).

Step 4: Use the forward time-differencing scheme to calculate the vorticity at the next time level, ζ^1.

Step 5: Calculate the streamfunction at the next time level, ζ^1, from (7.15) using over-relaxation (or another method for solving elliptic equations.)

Step 6: Enter the time loop, repeating Steps 3 through 5 as often as needed to advance the solution to the appropriate forecast time. However, for Step 4 within the time loop, use the leapfrog scheme instead of the forward scheme.

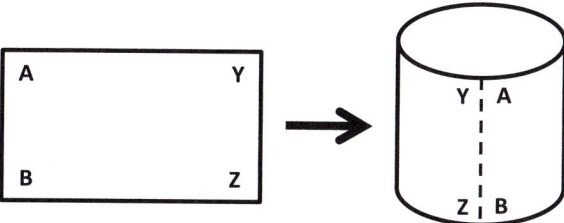

Fig. 7.3: For cyclic boundary conditions the left and right boundaries of the rectangular grid are brought together to form a cylinder.

7.6   Cyclic Boundary Conditions

Cyclic boundary conditions, also called periodic boundary conditions, allow the right-hand boundary of the grid to be placed adjacent to the left-hand boundary, so the grid essentially wraps around itself. This is illustrated in Fig. 7.3.

The relationship of the grid indexes near the boundaries is illustrated in Fig. 7.4 for a grid containing NX grid points in the x direction. The index of the first grid point (on the left boundary) is zero, while the index of the last grid point (on the right boundary) is $NX - 1$. Notice that the last grid point on the right boundary is not identical with the first grid point on the left boundary. They are still separated by a distance equal to the grid spacing.

When forming finite differences on the boundaries it is critical to use the proper indexes. As examples, consider calculating the meridional geostrophic wind from the streamfunction at the left and right boundaries using a centered difference scheme on a grid with

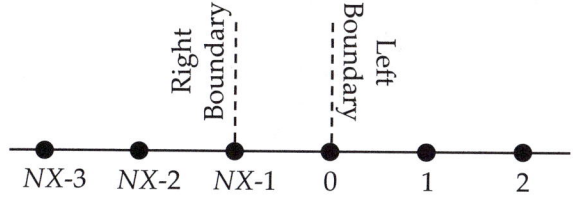

Fig. 7.4: Grid point notation near cyclic boundary for a grid with NX grid points.

spacing Δx. On the left boundary the appropriate expression is

$$v_0 = \frac{\psi_1 - \psi_{NX-1}}{2\Delta x} \tag{7.28}$$

while on the right boundary this would be written as

$$v_{N-1} = \frac{\psi_0 - \psi_{NX-2}}{2\Delta x}. \tag{7.29}$$

7.7   Including the Effects of Terrain

The effects of terrain can be incorporated into the barotropic model by including the Jacobian term involving the streamfunction and the terrain elevation. The leapfrog form of the prediction equation including terrain is

$$\zeta_{i,j}^{n+1} = \zeta_{i,j}^{n-1} - 2\Delta t \left([J(\psi, \zeta)]_{i,j}^n + \beta v_{i,j}^n + fgR_d T_s \left[J\left(\psi, z_s\right) \right]_{i,j}^n \right) \tag{7.30}$$

where T_s and z_s are the surface temperature and elevation. This can be turned into the equivalent-barotropic model by simply multiplying the last term by $\overline{A^2}$ to get

$$\zeta_{i,j}^{n+1} = \zeta_{i,j}^{n-1} - 2\Delta t \left([J(\psi, \zeta)]_{i,j}^n + \beta v_{i,j}^n + \overline{A^2} fgR_d T_s \left[J\left(\psi, z_s\right) \right]_{i,j}^n \right). \tag{7.31}$$

7.8   Baroclinic Filtered-Equation Model

A barotropic model is a single-level model. Since a primary mechanism for the development of cyclones in the atmosphere is baroclinic instability, which requires a vertical wind shear, a barotropic model cannot model cyclogenesis due to baroclinic instability. Instead, we need a model that has multiple vertical levels. A multi-level, QG baroclinic model can be constructed from the QG vorticity and QG thermodynamic energy equations discussed in the previous chapter and rewritten here,

$$\frac{\partial}{\partial t} \nabla^2 \psi + J(\psi, \nabla^2 \psi) + \beta \frac{\partial \psi}{\partial x} = f_0 \frac{\partial \omega}{\partial p} \tag{7.32}$$

$$\frac{\partial}{\partial t}\left(\frac{\partial \psi}{\partial p}\right) + J\left(\psi, \nabla \frac{\partial \psi}{\partial p}\right) + \frac{\sigma}{f_0}\omega = -\frac{R_d}{c_p f_0 p}\dot{Q}. \tag{7.33}$$

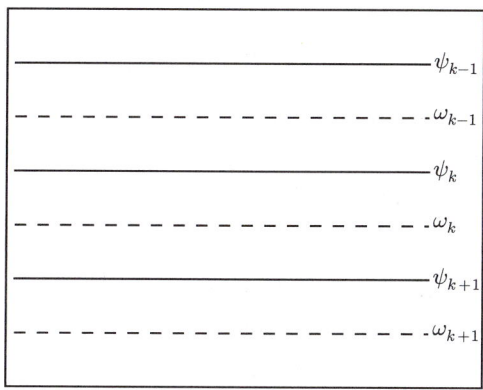

Fig. 7.5: Staggered vertical grid for QG baroclinic model. Streamfunction and vertical velocity are defined at alternate levels. Indexes increase downward since the vertical coordinate is pressure.

Equations (7.32) and (7.33) form a closed system of equations in two unknowns, ψ and ω (we assume that the heating rate, stability parameter, and all other values are known).

The concept of the baroclinic QG model is to apply (7.32) and (7.33) at multiple model levels. A staggered vertical grid is usually employed, with ψ and ω known at alternate levels (see Fig.7.5). The model equations at level k (writing only vertical finite differences) are

$$\frac{\partial}{\partial t}\nabla^2\psi_k + J(\psi_k, \nabla^2\psi_k) + \beta\frac{\partial\psi_k}{\partial x} = f_0\frac{\omega_k - \omega_{k-1}}{\Delta p} \qquad (7.34)$$

and

$$\omega_k = -\frac{f_0}{\sigma}\left\{ \begin{array}{l} \dfrac{\partial}{\partial t}\left(\dfrac{\psi_{k+1} - \psi_k}{\Delta p}\right) \\[2mm] -J\left[\dfrac{\psi_k + \psi_{k+1}}{2}, \nabla\left(\dfrac{\psi_{k+1} - \psi_k}{\Delta p}\right)\right] \\[2mm] -\dfrac{R_d}{c_p f_0 p}\dot{Q} \end{array} \right\}. \qquad (7.35)$$

Equation (7.34) is first solved at all levels to find the new values of ψ. Using these new values of ψ, (7.35) can then be solved for the new values of ω at all the levels. The process then repeats for each new time step.

Exercises

All of the exercises below will use a 2D, 101 × 51 domain (101 grid points in x-direction, 51 grid points in y-direction) and a grid spacing of 200 km. When doing the exercises be sure to convert to proper m·k·s units (meters, kilograms, seconds).

Ex. 7.1:
a. Create a streamfunction on the grid, given by the following formula

$$\psi_{i,j} = AJ + B + C\cos(ki) + D\exp\left(-\frac{r^2}{2\sigma^2}\right)$$

where

$$r = \sqrt{(i - 55.0)^2 + (j - 20.0)^2}$$

and the other parameters are: $A = -5.0 \times 10^6$ m^2 s^{-1}; $B = 7.4 \times 10^8$ m^2 s^{-1}; $C = 8.0 \times 10^7$ m^2 s^{-1}; $D = -1.5 \times 10^8$ m^2 s^{-1}; $k = \pi/25$; and $\sigma = 10.0$. When plotted the streamfunction should look like Figure 7.6.
b. Compute the Laplacian (vorticity) of the streamfunction from Ex. 7.1a, using the five-point stencil. Remember to use all m·k·s units in your calculations. When plotted the Laplacian should look like Figure 7.7. Note: The vorticity can only be calculated on the interior grid points. The vorticity on the boundaries should be initialized to zero.

Ex. 7.2: From the same streamfunction that you created in Ex. 7.1, calculate the Jacobian using (7.14). Remember to use all m·k·s units in your calculations. When plotted the result should look like Figure 7.8.
Note: The Jacobian cannot be calculated on grid points that are either on the boundary or adjacent to the boundary. There will be two rows or two columns along each boundary for which the Jacobian should be initialized to zero.

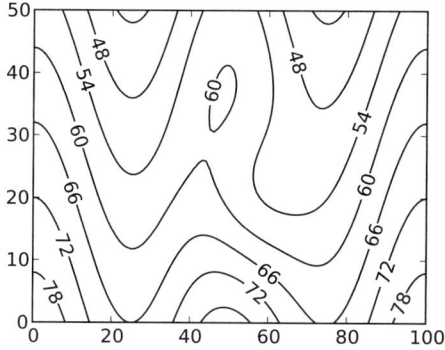

Fig. 7.6: Streamfunction for Ex. 7.1a. Units are $\times 10^{-7}$ m^2 s^{-1}.

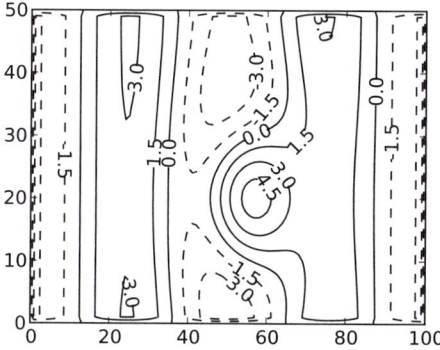

Fig. 7.7: Vorticity field for Ex. 7.1b. Units are $\times 10^5$ s^{-1}.

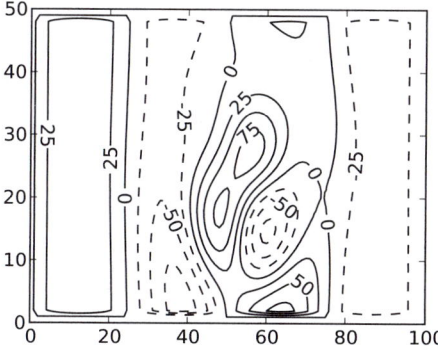

Fig. 7.8: Jacobian from Ex. 7.2. Units are $\times 10^{-11}$ s^{-2}.

Ex. 7.3:

a. Create a vorticity field on the grid, given by the formula

$$\zeta_{i,j} = A_1 \exp\left(-\frac{r_1^2}{2\sigma_1^2}\right) - A_2 \exp\left(-\frac{r_2^2}{2\sigma_2^2}\right)$$

where

$$A_1 = 7.0 \times 10^{-5}\ \text{s}^{-1}$$
$$A_2 = 4.9 \times 10^{-5}\ \text{s}^{-1}$$
$$\sigma_1 = 5.0$$
$$\sigma_2 = 7.5$$
$$r_1 = \sqrt{(i - 65.0)^2 + (j - 30.0)^2}$$
$$r_2 = \sqrt{(i - 30.0)^2 + (j - 20.0)^2}\ .$$

When plotted the vorticity field should look like Figure 7.9.

b. Starting with a guess field of zero everywhere on the grid, use the method of over-relaxation to find the streamfunction consistent with the vorticity field from Part a. Use a threshold value of $10^{-7}\ \text{s}^{-1}$. When plotted the final streamfunction should look like Figure 7.10.

c. Starting with a different guess field given by the formula

$$G_{i,j} = aj + b$$

where $a = -5.0 \times 10^6\ \text{m}^2\ \text{s}^{-1}$ and $b = 7.4 \times 10^8\ \text{m}^2\ \text{s}^{-1}$, use the method of over-relaxation to find the streamfunction consistent with the vorticity field from Part a. Use a threshold value of $10^{-7}\ \text{s}^{-1}$. The initial guess decreases linearly toward the North. When plotted the final streamfunction should look like Figure 7.11.

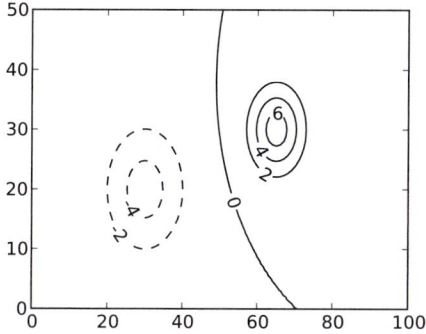

Fig. 7.9: Vorticity field for Ex. 7.3a. Units are $\times 10^5 \text{s}^{-1}$.

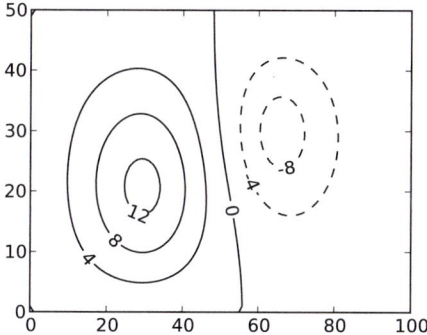

Fig. 7.10: Streamfunction for Ex. 7.3b. Units are $\times 10^{-7} \text{m}^2 \text{ s}^{-1}$.

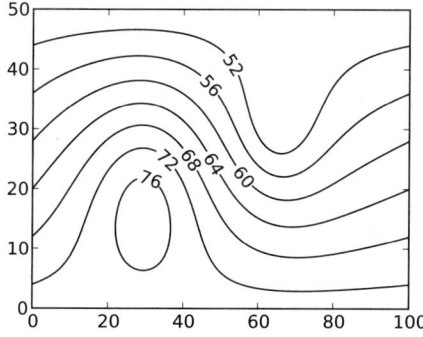

Fig. 7.11: Streamfunction for Ex. 7.3c. Units are $\times 10^{-7} \text{ m}^2 \text{ s}^{-1}$.

Ex. 7.4:

a. Create a streamfunction field on the grid, given by the formula

$$\psi_{i,j} = Aj + B + C \exp\left(-\frac{r_1^2}{2\sigma_1^2}\right) + D \exp\left(-\frac{r_2^2}{2\sigma_2^2}\right)$$

where

$$A = -5.0 \times 10^6 \text{ m}^2\text{s}^{-1}$$
$$B = 7.4 \times 10^8 \text{ m}^2\text{s}^{-1}$$
$$C = -6.0 \times 10^7 \text{ m}^2\text{s}^{-1}$$
$$D = 5.0 \times 10^7 \text{ m}^2\text{s}^{-1}$$
$$\sigma_1 = 3.5$$
$$\sigma_2 = 6.0$$
$$r_1 = \sqrt{(i - 25.0)^2 + (j - 20.0)^2}$$
$$r_2 = \sqrt{(i - 35.0)^2 + (j - 30.0)^2}$$

When plotted the final streamfunction should look like Figure 7.12.
b. Use the streamfunction created in Part a. as the initialization for a barotropic QG model. Use the following parameters:

- Δt: 15 minutes (900 s)

- relaxation threshold ε: 10^{-7} s^{-1}

- total time of simulation: 72 hours

- fixed boundaries

- write output every 6 hours

When plotted the model output at 24, 48, and 72 hours should look like Figure 7.13.

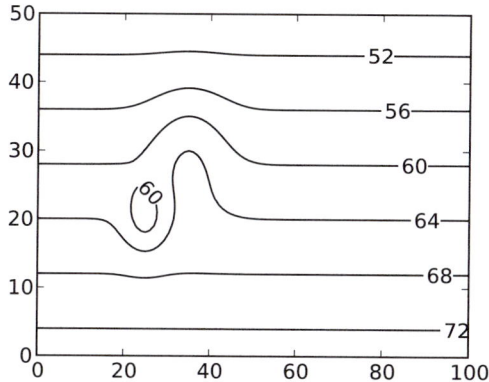

Fig. 7.12: Initial streamfunction for Ex. 7.4a. Units are $\times 10^{-7}$ m²s⁻¹.

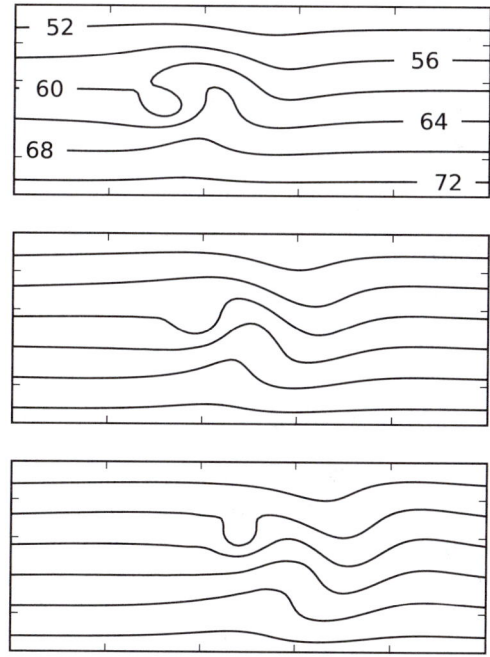

Fig. 7.13: Streamfunction for Ex. 7.4b at 24, 48, and 72 hours. Units are $\times 10^{-7}$ m² s⁻¹.

The following exercises are a bit more challenging.

Ex. 7.5:

a. Create a streamfunction field on the grid, given by the formula in Ex. 7.4, only change r_1 and r_2 to be

$$r_1 = \sqrt{(i - 70.0)^2 + (j - 20.0)^2}$$
$$r_2 = \sqrt{(i - 80.0)^2 + (j - 30.0)^2}$$

which shifts the disturbance toward the right boundary. When plotted the final streamfunction should look like Figure 7.14.

b. Use the streamfunction created in Part a. as the initialization for a barotropic QG model with the same parameters as in Ex. 7.4, but using cyclic boundary conditions. When plotted the model output at 24, 48, and 72 hours should look like Figure 7.15. Note that the disturbance moves off the right boundary and reappears on the left boundary.

Ex. 7.6: In this exercise we add a terrain term to the model from Ex. 7.5 in order to model a lee-side trough. We will assume a barotropic atmosphere with an isothermal temperature profile so that the terrain term in (7.30) is

$$f g R_d T_s \left[J \left(\psi, z_s \right) \right]_{i,j}^n .$$

For the parameter use: $f = 10^{-4} \ \text{s}^{-1}$, $g = 9.81 \ \text{m s}^{-2}$, and $T_s = 287 \ \text{K}$. For the model terrain use a Gaussian-shaped hill centered in the domain and given by the formula

$$z_{s(i,j)} = A \exp \left[-\frac{(i - 50)^2}{2\sigma_x^2} \right] \exp \left[-\frac{(i - 25)^2}{2\sigma_y^2} \right]$$

where $A = 1500$ meters, $\sigma_x = 4.0$, and $\sigma_y = 7.0$. When plotted the terrain should look like Figure 7.16.

We want to model lee-side troughing in a steady westerly flow, so the initial streamfunction will be defined using the formula

$$\psi_{i,j} = aj + b$$

where $a = -5.0 \times 10^6 \ \text{m}^2 \ \text{s}^{-1}$ and $b = 7.5 \times 10^8 \ \text{m}^2 \ \text{s}^{-1}$. The initial streamfunction when plotted should look like Figure 7.17.

The model output at 24, 48, and 72 hours should look like Figure 7.18.

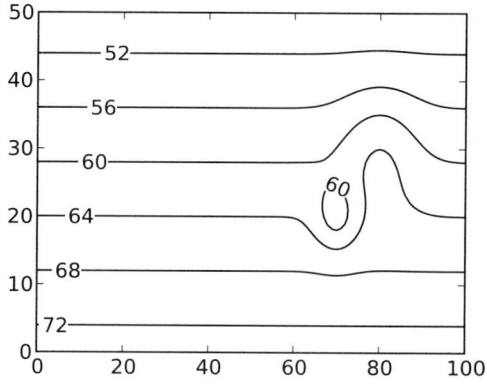

Fig. 7.14: Initial streamfunction for Ex. 7.5a. Units are $\times 10^{-7}$ m^{2}s^{-1}.

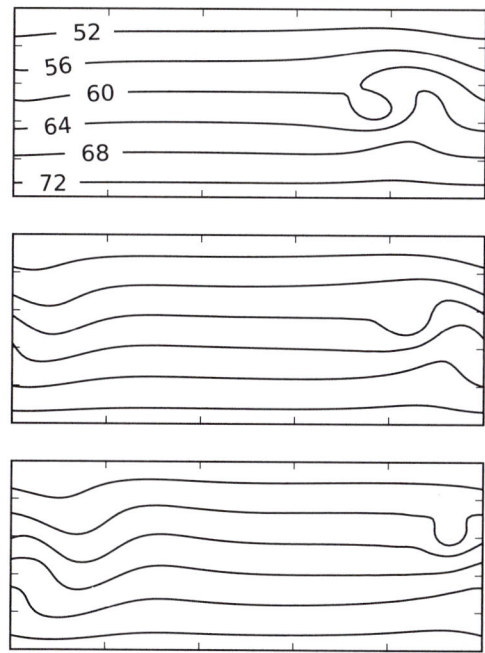

Fig. 7.15: Streamfunction for Ex. 7.5b at 24, 48, and 72 hours. Units are $\times 10^{-7}$ m^{2}s^{-1}.

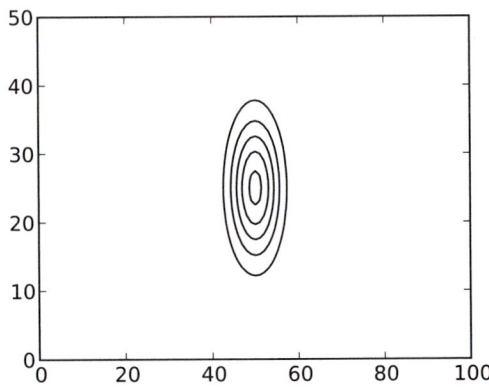

Fig. 7.16: Terrain for Ex. 7.6. Contours are 300, 600, 900, 1200, and 1500 meters.

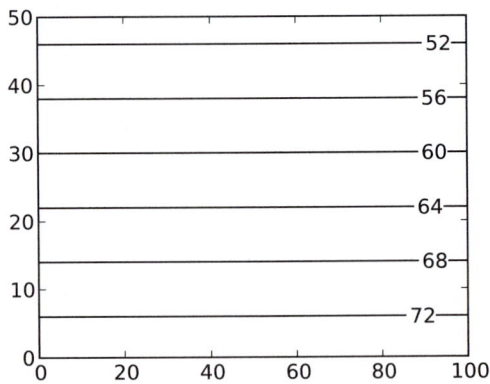

Fig. 7.17: Initial streamfunction for Ex. 7.6. Units are $\times 10^{-7}$ m^2 s^{-1}.

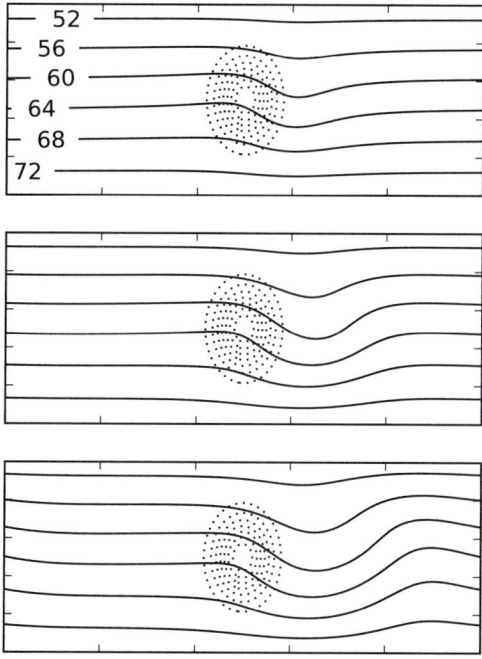

52
56
60
64
68
72

Fig. 7.18: Streamfunction for Ex. 7.6 at 24, 48, and 72 hours. Units are $\times 10^{-7}$ m^2 s^{-1}. Dashed lines are terrain elevation.

## CHAPTER 8

Barotropic Primitive Equation Models

8.1 Introduction

Until now we have illustrated the concepts of finite differencing with a single equation and one dependent variable. The moist atmosphere is governed by a system of seven equations with seven dependent variables. Jumping directly to the primitive equations would be rather daunting. Fortunately, there is a smaller and less complex subset of the primitive equations called the *barotropic primitive equations*,[1] which is ideal for illustrating many of the concepts needed for applying finite differencing to the full set of primitive equations. This subset describes the dynamics of a hydrostatic, constant-density fluid. Although the atmosphere is not a constant-density fluid, the barotropic primitive equations are readily extendable to large-scale flows in fluids consisting of multiple layers. They are not only useful for providing a simplified illustration of the numerical solution of a system of equations, but are applicable in their

---

[1]These equations are also known as the *shallow-water equations*, a name which disguises their importance for atmospheric modeling. Therefore we will use the name *barotropic primitive equations* in this book. Models based on these equations are known as barotropic primitive equation models, and are different from the barotropic QG models studied in the previous chapter.

own right as valid equations for use in certain operational and research models.

In this chapter the barotropic primitive equations are introduced and the dispersion properties of the waves supported by them are derived and discussed. The solution of the equations using numerical methods is then presented, leading to a discussion of the advantages of staggered versus unstaggered grids. The numerical stability conditions for these equations are also presented.

8.2   The Barotropic Primitive Equations

For a hydrostatic fluid of constant density ρ_1 lying over a flat bottom there are three governing equations,

$$\frac{\partial u}{\partial t} + u\frac{\partial u}{\partial x} + v\frac{\partial u}{\partial y} = -g'\frac{\partial \eta}{\partial x} + fv \tag{8.1}$$

$$\frac{\partial v}{\partial t} + u\frac{\partial v}{\partial x} + v\frac{\partial v}{\partial y} = -g'\frac{\partial \eta}{\partial y} - fu \tag{8.2}$$

$$\frac{\partial \eta}{\partial t} + u\frac{\partial \eta}{\partial x} + v\frac{\partial \eta}{\partial y} = -(H+\eta)\left(\frac{\partial u}{\partial x} + \frac{\partial v}{\partial y}\right) \tag{8.3}$$

where H is the mean depth (at rest) of the fluid and η is the displacement from the mean position of the interface between the fluid layer and whatever fluid lies above it (see Fig. 8.1). The term g' is called *reduced gravity*, and is given by

$$g' = \frac{\rho_1 - \rho_2}{\rho_1}g \tag{8.4}$$

where ρ_2 is the density of the fluid lying above the interface. Note that if the densities of the two fluids have widely disparate densities, such as air and water, then the reduced gravity is just equal to gravity. Equations (8.1) and (8.2) are the u- and v-momentum equations while (8.3) is the continuity equation. The three dependent variables are the u and v components of velocity and the displacement of the interface from equilibrium, η.

Equations (8.1) through (8.3) support the propagation of barotropic *inertio-gravity waves*, which are gravity waves influenced by the Earth's rotation. These are important for the adjustment

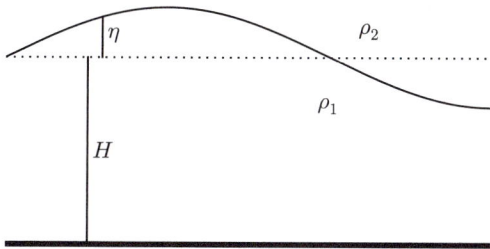

Fig. 8.1: Relationship of variables for the barotropic primitive equations. The mean (undisturbed) depth of the fluid is H, while the displacement of the interface from the mean is η. The densities of the lower and upper fluids are ρ_1 and ρ_2 respectively.

process by which the large-scale atmosphere and ocean maintain hydrostatic and geostrophic/gradient balance. Recall that these waves are 'filtered out' when the quasi-geostrophic approximation is made. So, unlike a QG model, a barotropic PE model can simulate the geostrophic/gradient adjustment process.

8.3 Waves

8.3.1 Properties of waves

This section presents a brief overview of wave properties and dispersion. A wave traveling in the positive x, y, and z directions is given by an equation of the form

$$u\left(x, y, z, t\right) = Ae^{\iota(kx+ly+mz-\omega t)},\tag{8.5}$$

where A is the complex amplitude of the wave. The parameters k, l, and m are the angular wave numbers in the x, y, and z directions, and are the components of the wave number vector, $\vec{K}$, which is given by

$$\vec{K} = k\hat{i} + l\hat{j} + m\hat{k}.\tag{8.6}$$

This vector points in the direction that the individual wave crests are traveling, and has a magnitude given by

$$K = \sqrt{k^2 + l^2 + m^2}.\tag{8.7}$$

The angular wave number and the wave length of the wave are related via

$$K = 1/\lambda. \tag{8.8}$$

The angular frequency of the wave is denoted by ω.

8.3.2   Wave dispersion

The *dispersion relation* for a particular type of wave is a mathematical equation that relates the frequency, ω, to the wave number components (k, l, and m) as well as to the physical properties of the system such as fluid depth, gravity, surface tension, Coriolis parameter, etc.,

$$\omega = F\left(k, l, m, [\text{physical parameters}]\right). \tag{8.9}$$

The *phase speed* is defined as the speed of progression of an individual wave crest, and is found by dividing the frequency by the magnitude of the wave number,

$$c = \omega/K. \tag{8.10}$$

The *phase velocity* is

$$\vec{c} = \frac{\omega}{K^2}(k\hat{i} + l\hat{j} + m\,\hat{k}). \tag{8.11}$$

The energy of a wave moves at the *group velocity*, and is given by,

$$\vec{c}_g = \frac{\partial \omega}{\partial k}\hat{i} + \frac{\partial \omega}{\partial l}\hat{j} + \frac{\partial \omega}{\partial m}\hat{k}. \tag{8.12}$$

If the phase speed of the waves does not depend on the wave number, then all of the waves, regardless of wave number, travel at the same speed. Such waves are said to be *nondispersive*. A signal comprised of nondispersive waves will maintain its shape with time. Another property of nondispersive waves is that the group velocity and the phase velocity are equal. If the phase speed depends on the wave number, then waves of different wave numbers will travel at different speeds and the waves are said to be *dispersive*. In a dispersive system the signal shape will change with time.

8.3.3 Linear waves

The primitive equations are *nonlinear* in that they include terms involving products of the dependent variables. The advection terms in particular are nonlinear. Nonlinear equations support nonlinear waves, which differ markedly from linear waves in that nonlinear waves may interact with one another and exchange energy between them, while linear waves do not interact with one another. Linear waves are also much easier to study since the mathematics is simpler. Many properties and aspects of real atmospheric waves can be effectively studied using a linearized set of equations. The process of linearizing the primitive equation is described in many texts on dynamic meteorology or in other sources. In this chapter we will use the linearized form of the barotropic primitive equations without deriving them.

8.3.4 Background flow

Waves in the atmosphere are usually embedded in a mean background flow. The mean flow modifies the frequency of linear waves via Doppler shifting, with the modified wave having frequency ω' given by

$$\omega' = \bar{u}k + \bar{v}l + \bar{w}m + \omega, \tag{8.13}$$

where $\bar{u}, \bar{v},$ and $\bar{w}$ are the $x, y,$ and z components of the background flow velocity. For linear waves the mean flow is simply added to the components of the phase and group velocities, with

$$\vec{c} = \left(\bar{u} + \frac{\omega k}{K^2}\right)\hat{i} + \left(\bar{v} + \frac{\omega l}{K^2}\right)\hat{j} + \left(\bar{w} + \frac{\omega m}{K^2}\right)\hat{k} \tag{8.14}$$

and

$$\vec{c}_g = \left(\bar{u} + \frac{\partial \omega}{\partial k}\right)\hat{i} + \left(\bar{v} + \frac{\partial \omega}{\partial l}\right)\hat{j} + \left(\bar{w} + \frac{\partial \omega}{\partial m}\right)\hat{k}. \tag{8.15}$$

8.4 Barotropic Gravity Waves

8.4.1 Dispersion properties

Since gravity waves are so important for the geostrophic/gradient adjustment process we need to understand their dispersion properties (dispersion relation, phase speed, group speed) so we can assess how these properties are altered when we approximate the barotropic primitive equations using finite differencing. In order to do this we must first find the dispersion relation for barotropic gravity waves.

The linearized versions of Eqs. (8.1) through (8.3) are

$$\frac{\partial u}{\partial t} + \bar{u}\frac{\partial u}{\partial x} + \bar{v}\frac{\partial u}{\partial y} = -g\frac{\partial \eta}{\partial x} + fv \qquad (8.16)$$

$$\frac{\partial v}{\partial t} + \bar{u}\frac{\partial v}{\partial x} + \bar{v}\frac{\partial v}{\partial y} = -g\frac{\partial \eta}{\partial y} - fu \qquad (8.17)$$

$$\frac{\partial \eta}{\partial t} + \bar{u}\frac{\partial \eta}{\partial x} + \bar{v}\frac{\partial \eta}{\partial y} = -H\left(\frac{\partial u}{\partial x} + \frac{\partial v}{\partial y}\right) \qquad (8.18)$$

where $\bar{u}$ and $\bar{v}$ are the components of the mean background flow. The terms involving $\bar{u}$ and $\bar{v}$ are called the *advection terms*. We have assumed that the density difference of the two fluids is large so that we can use gravity in place of reduced gravity in (8.16) and (8.17). To further simplify we will assume that the mean flow is zero, since the effects of the mean flow on dispersion can simply be added in later using (8.13). With these assumptions the equations simplify to

$$\frac{\partial u}{\partial t} = -g\frac{\partial \eta}{\partial x} + fv \qquad (8.19)$$

$$\frac{\partial v}{\partial t} = -g\frac{\partial \eta}{\partial y} - fu \qquad (8.20)$$

$$\frac{\partial \eta}{\partial t} = -H\left(\frac{\partial u}{\partial x} + \frac{\partial v}{\partial y}\right). \qquad (8.21)$$

To find the dispersion relation for waves supported by (8.19) through (8.21) we substitute into them the following 2-dimensional wavelike solutions for u, v, and η,

$$u(x,y,t) = Ae^{\iota(kx+ly-\omega t)} \qquad (8.22)$$

$$v(x,y,t) = Be^{\iota(kx+ly-\omega t)} \qquad (8.23)$$

$$\eta(x,y,t) = Ce^{\iota(kx+ly-\omega t)}, \qquad (8.24)$$

which results in the following three equations for the constant amplitudes A, B, and C

$$\begin{aligned}
\imath\omega A + fB - \imath kgC &= 0 \\
fA - \imath\omega B + \imath lgC &= 0 \\
kHA + lHB - \omega C &= 0
\end{aligned} \qquad (8.25)$$

(see Exercise 8.1). Equations (8.25) can be written in matrix form as

$$\begin{pmatrix} \imath\omega & f & -\imath kg \\ f & -\imath\omega & \imath lg \\ kH & lH & -\omega \end{pmatrix} \begin{pmatrix} A \\ B \\ C \end{pmatrix} = \begin{pmatrix} 0 \\ 0 \\ 0 \end{pmatrix}. \qquad (8.26)$$

In order for there to be a nontrivial solution for the wave amplitudes A, B, and C the determinant of the coefficient matrix must be zero,

$$\begin{vmatrix} \imath\omega & f & -\imath kg \\ f & -\imath\omega & \imath lg \\ kH & lH & -\omega \end{vmatrix} = 0. \qquad (8.27)$$

When (8.27) is multiplied out and solved for ω we obtain the dispersion relation

$$\omega = \pm\sqrt{gHK^2 + f^2}, \qquad (8.28)$$

where K is the total wave number given by

$$K = \sqrt{k^2 + l^2}. \qquad (8.29)$$

Equation (8.28) is the *dispersion relation* for linear barotropic *inertio-gravity waves*. From the dispersion relation the phase velocity for these waves is found to be

$$\vec{c} = \frac{\sqrt{gHK^2 + f^2}}{K^2} \left(\pm k\,\hat{\imath} \pm l\,\hat{\jmath}\right) \qquad (8.30)$$

and the group velocity is

$$\vec{c}_g = \frac{gH}{\sqrt{gHK^2 + f^2}} \left(\pm k\,\hat{\imath} \pm l\,\hat{\jmath}\right). \qquad (8.31)$$

Barotropic inertio-gravity waves are dispersive since their phase speed depends on the wave number (different wave numbers travel at different speeds.)

For short enough wave lengths (large enough wave numbers) the Coriolis term adds a negligible contribution to the dispersion relation. In this case, instead of barotropic inertio-gravity waves, we have ordinary barotropic gravity waves whose dispersion characteristics are

$$\omega = \pm\sqrt{gH}K \tag{8.32}$$

$$\vec{c} = \vec{c}_g = \frac{\sqrt{gH}}{K}\left(\pm k\hat{i} \pm l\hat{j}\right). \tag{8.33}$$

8.4.2 The Rossby radius of deformation

The *Rossby radius of deformation* (denoted as λ_R and often simply referred to as the radius of deformation) is the distance traveled by a gravity wave during one angular inertial period.[2] For a barotropic fluid the radius of deformation is given by the formula

$$\lambda_R = c/f, \tag{8.34}$$

where c is the speed of a barotropic gravity wave, which from (8.33) is

$$c = \sqrt{gH}. \tag{8.35}$$

The radius of deformation provides a length scale for determining the relative importance of the Earth's rotation. For a circulation whose characteristic length is much smaller than the radius of deformation, the Earth's rotation can be ignored as it is not important for the dynamics of the circulation.

8.4.3 Extension to multi-layered fluids

Gravity waves also exist in fluids comprised of multiple, constant-density layers. In a two-layer fluid, such as warm freshwater on top of cold salty water, there are two modes of oscillation. The first mode is a barotropic mode that behaves nearly identically to that which would occur if the fluid were comprised only of a single layer. The barotropic mode has a phase speed given by

$$c_0 = \sqrt{g\left(H_1 + H_2\right)} = \sqrt{gH} \tag{8.36}$$

---

[2]An angular inertial period has units of radians per second and is given by $1/f$.

where the subscript '0' indicates the barotropic mode and H is the total undisturbed depth of the fluid. The baroclinic mode is much slower, and has a phase speed given by

$$c_1 = \sqrt{gH_e},\qquad(8.37)$$

where H_e is called the *equivalent depth*, and is given by

$$H_e = \frac{(\rho_1 - \rho_2)}{\rho_1}\frac{H_1 H_2}{H},\qquad(8.38)$$

where ρ_1 and ρ_2 are the densities of the lower and upper layers, and H_1 and H_2 are the undisturbed depths of the lower and upper layers. A three-layer fluid would have three modes of oscillation: a single barotropic mode and two baroclinic modes. In general, an n-layered fluid would have n modes of oscillation, these being a single barotropic mode and $n - 1$ baroclinic modes. Each mode (denoted by i) has its own equivalent depth, $H_{e(i)}$, and wave speed given by

$$c_i = \sqrt{gH_{e(i)}}.\qquad(8.39)$$

Each mode also has its own radius of deformation given by

$$\lambda_{R(i)} = c_i/f.\qquad(8.40)$$

A continuously-stratified hydrostatic fluid can be thought of as having an infinite number of constant-density layers and as such has an infinite number of baroclinic modes, each with its own wave speed. The concept of equivalent depth therefore allows the barotropic primitive equations to be used for multi-layered, hydrostatic fluids, and has many applications in both atmospheric and oceanic modeling.

Since the baroclinic mode waves have speeds that are much smaller than the barotropic mode, the baroclinic radii of deformation are commensurately smaller than the barotropic radius of deformation. This means that the effects of the Earth's rotation may be important for small-scale baroclinic circulations, while for a barotropic circulation of same size, the effects of Coriolis could be negligible.

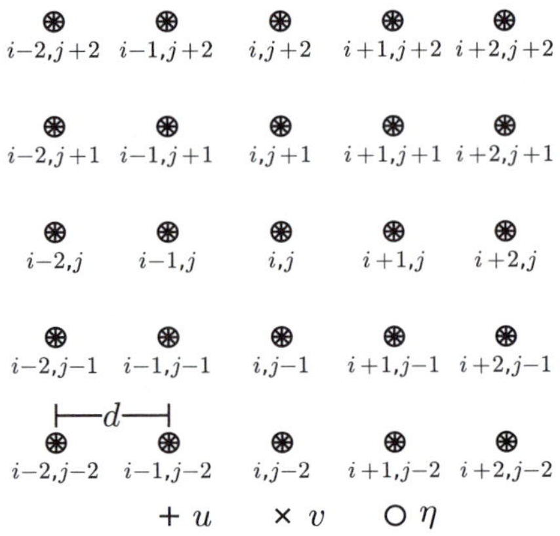

$$
\begin{array}{ccccc}
\circledast & \circledast & \circledast & \circledast & \circledast \\
i-2,j+2 & i-1,j+2 & i,j+2 & i+1,j+2 & i+2,j+2 \\[1em]
\circledast & \circledast & \circledast & \circledast & \circledast \\
i-2,j+1 & i-1,j+1 & i,j+1 & i+1,j+1 & i+2,j+1 \\[1em]
\circledast & \circledast & \circledast & \circledast & \circledast \\
i-2,j & i-1,j & i,j & i+1,j & i+2,j \\[1em]
\circledast & \circledast & \circledast & \circledast & \circledast \\
i-2,j-1 & i-1,j-1 & i,j-1 & i+1,j-1 & i+2,j-1 \\[1em]
\circledast & \circledast & \circledast & \circledast & \circledast \\
i-2,j-2 & i-1,j-2 & i,j-2 & i+1,j-2 & i+2,j-2
\end{array}
$$

$\vdash\!\!-d\!\!-\!\dashv$

$+\ u \qquad \times\ v \qquad \bigcirc\ \eta$

Fig. 8.2: Representation of a regular 2D grid in two dimensions. On this grid the u, v, and η values are all defined at every grid point. This grid is referred to as the *Arakawa A grid*. The grid spacing is d.

8.5   Regular 2D Grid

For modeling purposes we need to represent the barotropic primitive equations in finite-difference form on a two-dimensional grid. Figure 8.2 shows such a grid with uniform grid spacing d. This grid is referred to as the *Arakawa A grid*, or simply the 'A' grid. On this grid the values of u, v, and η are all defined at collocated grid points.

8.5.1   Finite-difference representation

The finite-difference forms of (8.1), (8.2), and (8.3) on the A grid using centered-in-time, centered-in-space differencing are

$$
u_{i,j}^{n+1} = u_{i,j}^{n-1} - \frac{\Delta t}{d} u_{i,j}^n (u_{i+1,j}^n - u_{i-1,j}^n) - \frac{\Delta t}{d} v_{i,j}^n (u_{i,j+1}^n - u_{i,j-1}^n)
$$
$$
- g' \frac{\Delta t}{d} (\eta_{i+1,j}^n - \eta_{i-1,j}^n) + 2\Delta t f v_{i,j}^n \quad (8.41)
$$

$$
v_{i,j}^{n+1} = v_{i,j}^{n-1} - \frac{\Delta t}{d} u_{i,j}^n (v_{i+1,j}^n - v_{i-1,j}^n) - \frac{\Delta t}{d} v_{i,j}^n (v_{i,j+1}^n - v_{i,j-1}^n)
$$
$$
- g' \frac{\Delta t}{d} (\eta_{i,j+1}^n - \eta_{i,j-1}^n) - 2\Delta t f u_{i,j}^n \quad (8.42)
$$

$$\eta_{i,j}^{n+1} = \eta_{i,j}^{n-1} - \frac{\Delta t}{d} u_{i,j}^n (\eta_{i+1,j}^n - \eta_{i-1,j}^n) - \frac{\Delta t}{d} v_{i,j}^n (\eta_{i,j+1}^n - \eta_{i,j-1}^n)$$

$$- (H + \eta_{i,j}^n) \frac{\Delta t}{d} (u_{i+1,j}^n - u_{i-1,j}^n + v_{i,j+1}^n - v_{i,j-1}^n). \quad (8.43)$$

Although (8.41) through (8.43) may appear cumbersome, the concept of using them to predict the future values of u, v, and η is quite simple. At each grid point (i, j) all the values on the right-hand side of the equations are known, and so the new values of u, v, and η at time level $n + 1$ are able to be determined. Equations (8.41) through (8.43) are simply iterated as many times as needed to advance the solution to the desired future time.

8.5.2 Gravity wave dispersion

Since gravity waves are important for the geostrophic/gradient adjustment process we would like to evaluate how well these waves can be simulated on a regular 2D grid. Our hope is that when we represent the barotropic primitive equations in finite-difference form, the dispersion properties of the waves are not significantly different than those of true barotropic gravity waves. We find the dispersion relation for waves on the grid by first representing the linearized barotropic primitive equations, (8.19) through (8.21), using centered-in-space differencing. These take the form

$$\frac{\partial u}{\partial t} = -\frac{g}{2d}(\eta_{i+1,j} - \eta_{i-1,j}) + f v_{i,j} \quad (8.44)$$

$$\frac{\partial v}{\partial t} = -\frac{g}{2d}(\eta_{i,j+1} - \eta_{i,j-1}) - f u_{i,j} \quad (8.45)$$

$$\frac{\partial \eta}{\partial t} = -\frac{H}{2d}(u_{i+1,j} - u_{i-1,j} + v_{i,j+1} - v_{i,j-1}). \quad (8.46)$$

Equations (8.44) through (8.46) are known as *differential-difference equations*, since we have used finite differencing only for the spatial derivatives while leaving the time derivatives intact. We then assume discrete wavelike solutions of the form

$$u(i, j, t) = A e^{\iota(kid + ljd - \omega t)} \quad (8.47)$$

$$v(i, j, t) = B e^{\iota(kid + ljd - \omega t)} \quad (8.48)$$

$$\eta(i, j, t) = C e^{\iota(kid + ljd - \omega t)} \quad (8.49)$$

and substitute them into (8.44) through (8.46), resulting in the matrix equation

$$
\begin{pmatrix}
\iota\omega & f & -\frac{g}{2d}(e^{\iota kd} - e^{-\iota kd}) \\
f & -\iota\omega & \frac{g}{2d}(e^{\iota ld} - e^{-\iota ld}) \\
\frac{H}{2d}(e^{\iota kd} - e^{-\iota kd}) & \frac{H}{2d}(e^{\iota ld} - e^{-\iota ld}) & -\iota\omega
\end{pmatrix}
\begin{pmatrix} A \\ B \\ C \end{pmatrix}
=
\begin{pmatrix} 0 \\ 0 \\ 0 \end{pmatrix}. \quad (8.50)
$$

Setting the determinant of the coefficient matrix from (8.50) equal to zero and solving for ω results in the following dispersion relation for barotropic inertio-gravity waves on the A grid,

$$
\omega_A^2 = \frac{gH}{d^2}\left(\sin^2 kd + \sin^2 ld\right) + f^2. \quad (8.51)
$$

By dividing (8.51) by f^2 and using the definition of the radius of deformation from (8.34) the result is

$$
\omega_A^2/f^2 = \frac{\lambda_R^2}{d^2}\left(\sin^2 kd + \sin^2 ld\right) + 1. \quad (8.52)
$$

The true dispersion relation for barotropic inertio-gravity waves from (8.28) can be rewritten as

$$
\omega^2/f^2 = \frac{\lambda_R^2}{d^2}\left(k^2 d^2 + l^2 d^2\right) + 1. \quad (8.53)
$$

A plot of the true dispersion relation (8.53) is presented in Fig. 8.3, while Fig. 8.4 shows a plot of the dispersion relation on the A grid, (8.52). Significant differences between the two figures are readily apparent. Note that in Fig. 8.3 the contours of frequency are concentric rings around the origin. This is because dispersion of true barotropic inertio-gravity waves is isotropic, with no dependence on the direction the waves are traveling. A wave of a given wave number will have the same frequency regardless of whether it is traveling toward the North, East, South, West, or any direction in between. However, the dispersion on the A grid does depend on the direction of the wave motion.

Another significant difference between the two figures is that for a given wave number the waves on the A grid have a lower frequency (and therefore slower speed) than the actual waves. Even more troubling is that while the frequency of the actual waves is

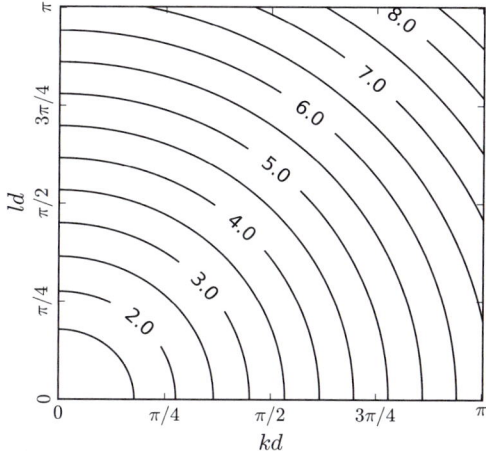

Fig. 8.3: Dispersion of true 2D barotropic inertio-gravity waves. Contours are $|\omega|/f$ for $\lambda_R/d = 2$.

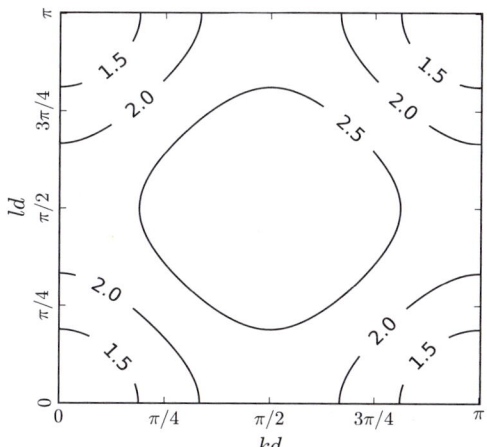

Fig. 8.4: Dispersion of 2D barotropic inertio-gravity waves on the Arakawa A grid. Contours are $|\omega|/f$ for $\lambda_R/d = 2$.

monotonically increasing with wave number, the frequency of the waves on the A grid actually reaches a maximum value and then begins to decrease as wave number increases. This is illustrated in Fig. 8.5, where values of $|\omega|/f$ are plotted and compared for both $l = 0$ and for $l = k$ (a line at 45° to the k axis and through the origin of the grid.) That the frequency reaches a maximum on the A grid and then decreases is a serious drawback, since this implies that the group speed is zero at the wave number corresponding to the maximum frequency, and that the group speed actually changes sign and becomes negative at wave numbers higher than where the maximum frequency occurs. Since energy propagates at the group speed, this means that on the A grid the wave energy may actually be traveling in the opposite direction than that of the true waves.

8.6   Staggered Grids

8.6.1   Configurations for staggered grids

The Arakawa A grid is not the only option for a 2D grid. There are many *staggered grids* that can be formulated by not defining the dependent variables at collocated grid points. Figures 8.6 through 8.9 illustrate four possibilities for staggered grids. These are referred to as the *Arakawa B, C, D,* and *E* grids.

On a staggered grid the grid spacing is always defined as the shortest distance between adjacent grid points of the same variable, and may actually be measured diagonally. Each variable has its own grid-index numbering system. As with the unstaggered grid, the shortest resolvable wave on the staggered grids also varies depending on the direction of propagation. For the B, C, and D grids the shortest resolvable wavelength in the x and y directions is $2d$ and in the diagonal direction it is $\sqrt{2}d$. For the E grid these values are flipped, with shorter waves resolved in the x and y directions.

Writing the finite-difference form of the equations on a staggered grid requires care and patience, and the resulting equations are more cumbersome than for an unstaggered grid. For example,

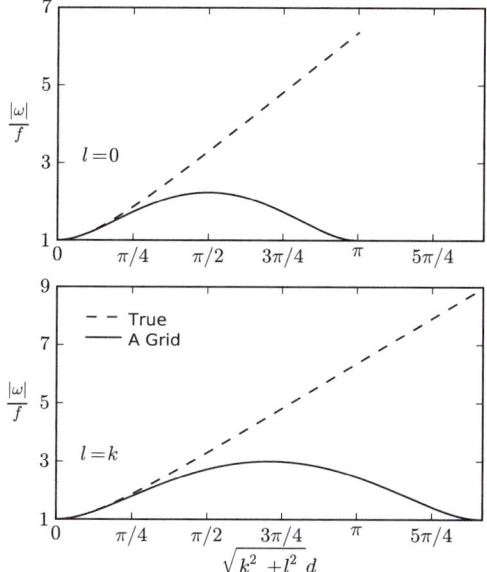

Fig. 8.5: Dispersion of barotropic inertio-gravity waves on the Arakawa A grid compared to the true solution. Contours are $|\omega|/f$ for $\lambda_R/d = 2$. The top panel is for waves traveling purely in the x direction, while the bottom panel is for diagonally propagating waves. Note that the group speed on the A grid is zero at the point of maximum frequency, and that the sign of the group speed switches as you move from a lower wave number to a higher wave number across the maximum.

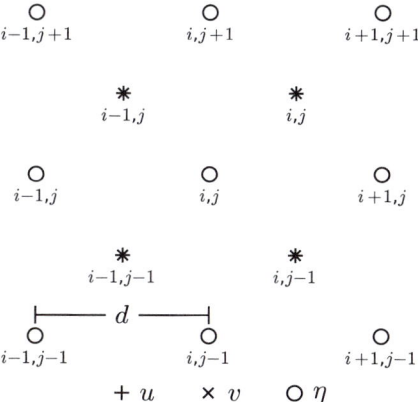

Fig. 8.6: Configuration for the Arakawa B grid. In this grid the u and v velocity components are collocated, but are separate from the η values.

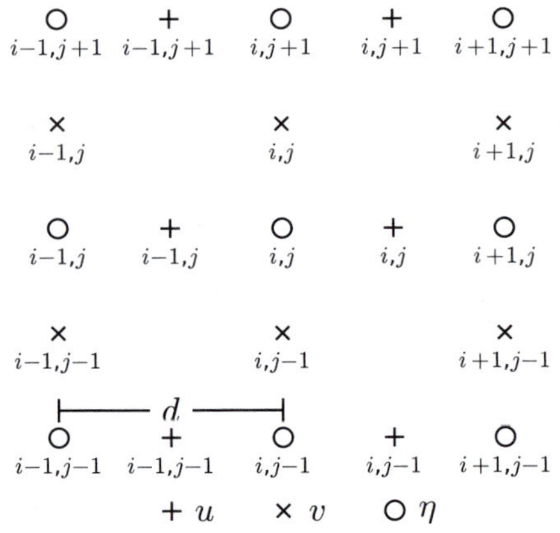

Fig. 8.7: Configuration for the Arakawa C grid. In this grid none of the values of u, v, or η values are collocated.

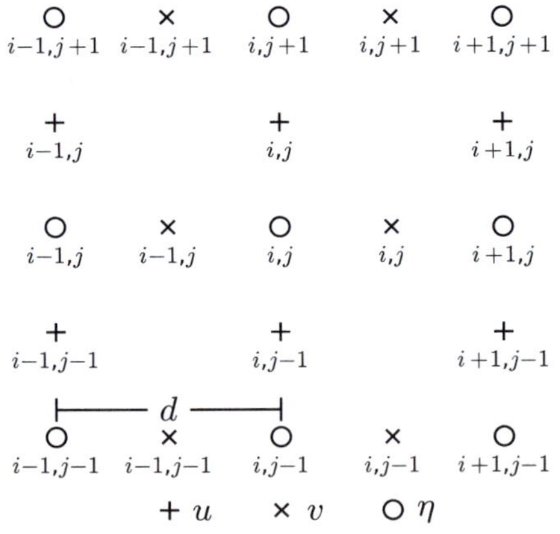

Fig. 8.8: Configuration for the Arakawa D grid. This is the C grid rotated $90°$ so that the u and v grid points are transposed.

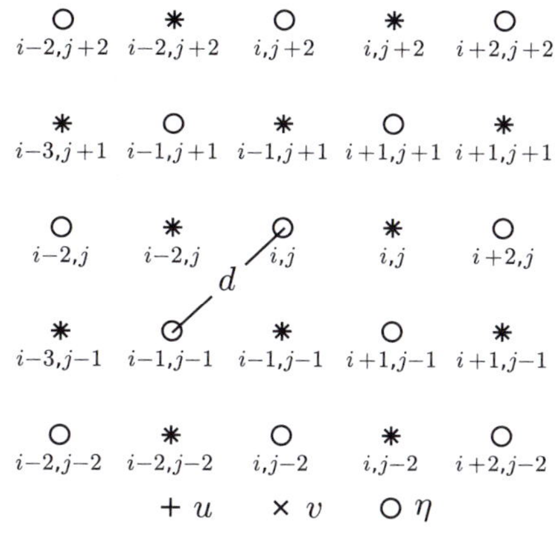

Fig. 8.9: Configuration for the Arakawa E grid. This is just the B grid rotated by 45°, or it can also be thought of as two C grids superimposed. The grid spacing on the E grid is measured diagonally.

on the C grid the u-momentum equation in finite-difference form is

$$\frac{u_{i,j}^{n+1} - u_{i,j}^{n-1}}{2\Delta t} + u_{i,j}^n \frac{\left(u_{i+1,j}^n - u_{i-1,j}^n\right)}{2d}$$

$$+ \frac{\left(v_{i,j}^n + v_{i+1,j}^n + v_{i,j-1}^n + v_{i+1,j-1}^n\right)}{4} \frac{\left(u_{i,j+1}^n - u_{i-1,j}^n\right)}{2d}$$

$$= -g\frac{\left(\eta_{i+1,j}^n + \eta_{i,j}^n\right)}{d} + \frac{f}{4}\left(v_{i,j}^n + v_{i+1,j}^n + v_{i,j-1}^n + v_{i+1,j-1}^n\right) \quad . \quad (8.54)$$

Note that since v is not defined at the same grid points were u is defined, then in the terms where v is needed we must average the values of v at the four surrounding grid points. Note also that when using a centered difference for the spatial derivative of u, we divide by $2d$, while the spatial derivative of η in the equation is found by dividing by d. Despite these drawbacks, staggered grids have enormous advantages compared to unstaggered grids, as will be illustrated in the next section.

8.6.2 Wave dispersion on staggered grids

The dispersion properties for waves on the staggered grids are derived in the same manner as those of the unstaggered grids. The

linearized equations are written as differential-difference equations
for each grid, and the solutions (8.47) through (8.49) are then substituted into them. The resulting set of algebraic equations for the
amplitudes A, B, and C are written as a matrix equation, the determinant set equal to zero, and the result solved for the frequency ω.
These derivations are left as exercises, but the results are presented
here. The dispersion relations for the B, C, D, and E grids are

$$\omega_B^2/f^2 = \frac{2\lambda_R^2}{d^2} \left(1 - \cos kd \cos ld\right) + 1 \tag{8.55}$$

$$\omega_C^2/f^2 = \frac{2\lambda_R^2}{d^2} \left[(1 - \cos kd) + (1 - \cos ld)\right]$$
$$+ \frac{1}{4} \left(1 + \cos kd\right) \left(1 + \cos ld\right) \tag{8.56}$$

$$\omega_D^2/f^2 = \frac{\lambda_R^2}{2d^2} \left[\sin^2 kd \left(1 + \cos ld\right) + \sin^2 ld \left(1 + \cos kd\right)\right]$$
$$+ \frac{1}{4} \left(1 + \cos kd\right) \left(1 + \cos ld\right) \tag{8.57}$$

$$\omega_E^2/f^2 = \frac{\lambda_R^2}{d^2} \left[2 - \cos(\sqrt{2}\,kd) - \cos(\sqrt{2}\,ld)\right] + 1. \tag{8.58}$$

These dispersion relations, along with that for the A grid, are plotted and compared with the true dispersion relation in Fig. 8.10. The
top panel of this figure shows that for waves traveling purely in the
x direction, the B and C grids have the best dispersion properties,
while the D and A grids are the worst. The bottom panel shows that
for diagonally propagating waves, the C grid has the best dispersion
properties, followed next by the E grid. For diagonally propagating
waves the D grid is still the worst, while the A and B grids have
identical dispersion properties. The rotational symmetry of the B
and E grids is also seen in Fig. 8.10, with the B-plot in the top panel
being identical to the E-plot in the bottom panel, and vice-versa.

Based on this dispersion analysis we conclude that the C grid is
overall a very good choice for a staggered 2D grid, and that the A
and D grids are very poor choices. In fact, the A and D grids are not
commonly used in numerical models of the atmosphere or ocean,

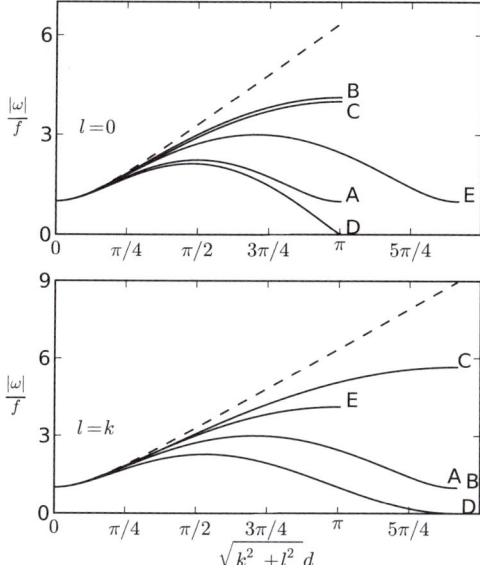

Fig. 8.10: Dispersion of barotropic inertio-gravity waves on the Arakawa A, B, C, D, and E grids compared to the true solution (dashed line). Contours are $|\omega|/f$ for $\lambda_R/d = 2$. The top panel is for waves traveling purely in the x direction, while the bottom panel is for diagonally propagating waves.

while the C grid is very commonly used. Though the B and E grids (which are essentially identical other than a 45° rotation) may not appear to perform well, it turns out that they are also frequently used in modeling for reasons that are discussed in the next section.

8.6.3 Static stability and grid selection

Although the concept of staggered grids has been illustrated using the barotropic primitive equations, the results are directly applicable to the full primitive equations. We simply substitute pressure in place of η in Figs. 8.6 through 8.9. We can also assess the effects of static stability on the dispersion properties of the grids. The Rossby radius of deformation in a baroclinic fluid is linked to the static stability of the fluid, with lower stability associated with a smaller radius of deformation for the dominant baroclinic modes. To see the effects of static stability on the dispersion properties of the grids, we replot Fig. 8.10, only for a much lower value of λ_R/f in order

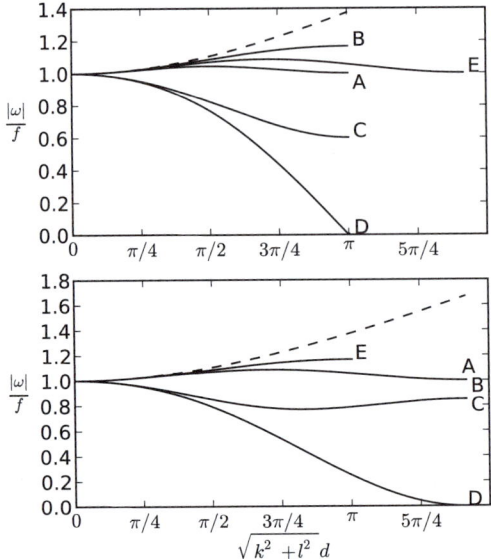

Fig. 8.11: Same as in Fig. 8.10 only for $\lambda_R/d = 0.3$, which represents a lower static stability environment. In this case, both the B and E grid outperform the C grid.

to represent a lower static stability environment. This plot is shown in Fig. 8.11 for a value of $\lambda_R/f = 0.3$. In this lower stability environment the dispersion properties of the B and E grids are actually better than the C grid. Model developers choose between the B, C, and E grids depending on the primary conditions for which the model will be developed. The C grid is the most common grid used in atmospheric models, but the B and E grids are also used, particularly for mesoscale models in highly baroclinic environments, in which the Rossby radius of deformation is small.

8.7 Numerical Stability for Barotropic PE Models

In Chapter 4 we derived the numerical stability criteria for various time and space schemes applied to the advection equation. We found that for the unfiltered leapfrog scheme the stability condition is

$$\left| \frac{c\Delta t}{d'} \right| \leq 1 \tag{8.59}$$

where d' is the effective grid spacing. We also discovered that the numerical stability requirement depends on both the equation being solved and on the time and space differencing chosen. In this section we explore the numerical stability of the leapfrog time-differencing scheme applied to the barotropic primitive equations on the A, B, C, and E grids.[3] We will also discuss the stability conditions for the 3rd-order Adams-Bashforth scheme.

In preparation for analyzing the stability of specific finite-difference schemes applied to the barotropic primitive equations, we will first prove that we can ignore the advection terms when deriving the stability conditions, as these can simply be added in at the end in a similar manner as was done for the dispersion relations. We will also not consider the Coriolis terms, and in Section 8.7.4 we will justify our ignoring these terms when assessing numerical stability.

8.7.1   The advection terms and stability

Here we will show that the effects of the advection terms on stability can simply be added to the stability requirement derived without the advection terms. This is illustrated most simply by using the one-dimensional linearized barotropic primitive equations without the Coriolis terms, which are

$$\frac{\partial u}{\partial t} + \bar{u}\frac{\partial u}{\partial x} = -g\frac{\partial \eta}{\partial x} \tag{8.60}$$

$$\frac{\partial \eta}{\partial t} + \bar{u}\frac{\partial \eta}{\partial x} = -H\frac{\partial u}{\partial x}. \tag{8.61}$$

In Exercise 8.3 we show that these two equations can be combined to form a single advection equation having the form of either

$$\frac{\partial}{\partial t}\left(u + \sqrt{g/H}\eta\right) + \left(\bar{u} + \sqrt{gH}\right)\frac{\partial}{\partial x}\left(u + \sqrt{g/H}\eta\right) = 0 \tag{8.62}$$

or

$$\frac{\partial}{\partial t}\left(u - \sqrt{g/H}\eta\right) + \left(\bar{u} - \sqrt{gH}\right)\frac{\partial}{\partial x}\left(u - \sqrt{g/H}\eta\right) = 0. \tag{8.63}$$

---

[3]The D grid is omitted here due to its poor dispersion properties. The A grid also has poor dispersion properties, but is retained for comparison with an unstaggered grid.

Equations (8.62) and (8.63) are both 1D advection equations of the form

$$\frac{\partial S}{\partial t} + c' \frac{\partial S}{\partial x} = 0 \qquad (8.64)$$

where

$$c' = \bar{u} \pm \sqrt{gH} = \bar{u} \pm c. \qquad (8.65)$$

We previously established that the stability condition for the 1D advection equation using the leapfrog scheme and centered differencing is

$$\left| \frac{c' \Delta t}{d} \right| \leq 1. \qquad (8.66)$$

The stability condition for either equation (8.60) or (8.61), using the leapfrog scheme and centered differencing, is therefore

$$\left| (\bar{u} \pm c) \frac{\Delta t}{d} \right| \leq 1. \qquad (8.67)$$

If $\bar{u}$ were zero the condition would simply be

$$\left| c \frac{\Delta t}{d} \right| \leq 1. \qquad (8.68)$$

These results show that the effect of the advection term on numerical stability is merely to add the advection speed $\bar{u}$ to the wave speed c in the stability condition. Therefore, in the following sections we omit the advection terms in the stability analysis, recognizing that if needed we can simply add them back into the derived stability condition.

8.7.2   Leapfrog scheme on the Arakawa C grid

We illustrate how the stability conditions for a particular time-scheme on a given grid are determined by using the C grid as an example. As always, we first need to find the amplification factors, which are found by writing the finite-difference forms of (8.19) through (8.21) for the chosen grid. For the leapfrog scheme and centered-space differencing on the C grid these are

$$u_{i,j}^{n+1} = u_{i,j}^{n-1} - \frac{2g\Delta t}{d}\left(\eta_{i+1,j}^n - \eta_{i,j}^n\right) \tag{8.69}$$

$$v_{i,j}^{n+1} = v_{i,j}^{n-1} - \frac{2g\Delta t}{d}\left(\eta_{i,j+1}^n - \eta_{i,j}^n\right) \tag{8.70}$$

$$\eta_{i,j}^{n+1} = \eta_{i,j}^{n-1} - \frac{2H\Delta t}{d}\left(u_{i,j}^n - u_{i-1,j}^n + v_{i,j}^n - v_{i,j-1}^n\right). \tag{8.71}$$

We then assume solutions of the form

$$u_j^n = \lambda^n A e^{\iota(kid+ljd)} \tag{8.72}$$

$$v_j^n = \lambda^n B e^{\iota(kid+ljd)} \tag{8.73}$$

$$\eta_j^n = \lambda^n C e^{\iota(kid+ljd)} \tag{8.74}$$

where λ is the amplification factor.[4] We substitute these solutions into the finite-difference equations (8.69) through (8.71), which results in the following matrix equation

$$\begin{pmatrix} (\lambda^2-1) & 0 & -\frac{2g\Delta t}{d}(1-e^{\iota kd})\lambda \\ 0 & (\lambda^2-1) & -\frac{2g\Delta t}{d}(1-e^{\iota ld})\lambda \\ \frac{2H\Delta t}{d}(1-e^{-\iota kd})\lambda & \frac{2H\Delta t}{d}(1-e^{-\iota ld})\lambda & (\lambda^2-1) \end{pmatrix} \begin{pmatrix} A \\ B \\ C \end{pmatrix} = \begin{pmatrix} 0 \\ 0 \\ 0 \end{pmatrix}. \tag{8.75}$$

For there to be nontrivial solutions for the amplitudes A, B, and C, the determinant of the coefficient matrix must be equal to zero. This results in the following equation for λ,

$$\left(\lambda^2 - 1\right)^2 + \sigma^2\lambda^2 = 0, \tag{8.76}$$

where

$$\sigma^2 = \frac{8c^2\Delta t^2}{d^2}\left(2 - \cos kd - \cos ld\right). \tag{8.77}$$

In (8.77) c is the speed of a barotropic gravity wave, $\sqrt{gH}$. Analysis of (8.76) (see Exercise 8.5) shows that the magnitude of the amplification factors will be less than one for values of σ meeting the condition

$$\sigma^2 \leq 4. \tag{8.78}$$

---

[4]As in Chapter 4, the subscript n on λ^n is a true exponent as well as a time-level indicator.

Table 8.1: Comparison of stability conditions for the leapfrog scheme on the A, B, C, and E grids. Criteria are reported in terms of the actual grid spacing d and the effective grid spacing d'.

	A Grid	B/E Grid	C Grid
$|c\Delta t/d| \leq$	$\sqrt{2}/2$	$1/2$	$\sqrt{2}/4$
$|c\Delta t/d'| \leq$	1	$\sqrt{2}/2$	$1/2$

The most limiting case is when $\cos kd = \cos ld = -1$, which means that in order to guarantee stability for all possible wave numbers, the stability condition

$$\left|\frac{c\Delta t}{d}\right| \leq \frac{\sqrt{2}}{4} \tag{8.79}$$

must hold. In terms of the effective grid spacing this condition is

$$\left|\frac{c\Delta t}{d'}\right| \leq \frac{1}{2}. \tag{8.80}$$

8.7.3   Stability compared among the grids

Derivation of the stability criteria for the leapfrog scheme applied to the other grids (A, B, and E) is left for the exercises. The results are compared in Table 8.1, and show that for the leapfrog scheme the C grid has the most stringent stability criterion, while the A grid has the least stringent.

In the exercises we also derive the stability condition for the 3rd-order Adams-Bashforth scheme on the B/E and C grids.[5] These results are somewhat reversed from the leapfrog scheme, with the B/E grid having the most stringent condition of

$$\left|\frac{c\Delta t}{d}\right| \leq 0.18 \tag{8.81}$$

while on the C grid the condition is

$$\left|\frac{c\Delta t}{d}\right| \leq 0.25. \tag{8.82}$$

---

[5]We do not bother computing this for the A and D grids simply because they are not commonly used.

8.7.4   Effects of the Coriolis term

We now address the effects of including the Coriolis terms in the stability analysis. To do this we analyze the stability condition for inertio-gravity waves traveling without advection on a simple 1D staggered grid where the u and v velocity components are collocated at grid points that are separate from the grid points where η is defined (see Exercise 8.6.) The resulting stability condition is

$$\left| \sqrt{\frac{c^2 \Delta t^2}{d^2} + \frac{f^2 \Delta t^2}{4}} \right| \leq \frac{1}{2} \qquad (8.83)$$

which can be rewritten in terms of the Rossby radius of deformation as

$$\left| f \Delta t \sqrt{\frac{\lambda_R^2}{d^2} + \frac{1}{4}} \right| \leq \frac{1}{2}. \qquad (8.84)$$

Inspection of (8.84) reveals that as long as λ_R^2/d^2 is much larger than 0.25, the Coriolis term plays a minimal effect in the stability condition. It is only for grid spacings that are large compared to the radius of deformation that the Coriolis terms become important in the stability analysis.

8.8   Two-interval Noise

The A grid can be thought of as being constructed of four superimposed C grids, while the B and E grids are constructed of two superimposed C grids. If leapfrog time differencing is used, then numerically these C grids are independent of each other because the predicted future values of the variables at grid points on one of these C grids does not depend on variables defined on grid points on the other C grids. This poses problems because there will be an independent solution on each C grid, and it is possible for these solutions to diverge from each other instead of remaining consistent. At its most extreme, if the solution on one of the C grids differed from the solution on another C grid by a constant value, a false stationary wave of wavelength $2d$ (wave number π/d) would be present throughout the grid. This wave is known as the *two-interval wave* or *two-interval noise* because of its wavelength being equal to twice the grid interval.

The two interval wave only occurs with leapfrog time differencing on the A, B, and E grids. It does not occur on the C grid, or with the 3rd-order Adams-Bashforth scheme on any grid (another reason to prefer this scheme over the leapfrog scheme). If leapfrog time differencing is used on the A, B, or E grids, then additional measures must be applied to suppress the two-interval wave. Such measures could be filtering or occasional application of a time-differencing scheme that does not support the two-interval wave. As newer computing technology eases the constraints imposed on computer storage and speed, the use of higher-order time-differencing schemes such as the 3rd-order Adams-Bashforth schemes is more common, and the two-interval wave created by using the leapfrog scheme is less of an issue for modelers.

8.9   Time-staggered Grids

For completeness we mention that it is also possible to use grids that are staggered in time as well as in space. As a simplistic example consider the 1D staggered grid shown in Fig. 8.12 applied to the barotropic primitive equations with no mean flow or Coriolis terms. The vertical axis depicts the different time levels, $n - 2$ through $n + 2$. If centered-in-space differencing is used in conjunction with leapfrog time differencing, then the variables at the grid points connected by the dotted lines are completely independent from the grid points that are not connected by the dotted lines. There are essentially two independent subgrids. If only one of these subgrids is retained, the result is a grid that is both staggered-in-time and staggered-in-space and requiring half the computational effort for a solution as the original grid.

A possible two-dimensional grid that is staggered both in time and in space consists of two independent Arakawa D grids, one applied at the even time levels and one applied at the odd time levels, as shown in Fig. 8.13. This configuration is called the *Eliassen grid*. Even though a single Arakawa D grid that is unstaggered in time would have very poor qualities, when two such D grids that are staggered in-time are used, their dispersion properties are similar to that of an E grid.

Use of time staggering is only appropriate when leapfrog time differencing scheme is used. For other time differencing

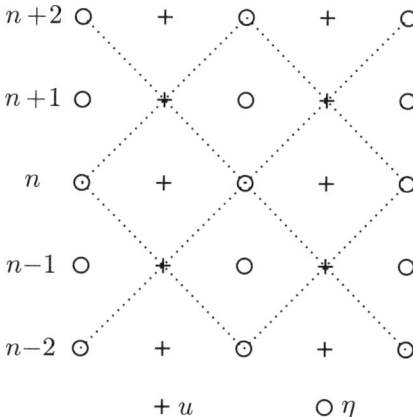

Fig. 8.12: A 1D staggered-in-space grid showing the different time levels. If leapfrog time differencing is used with centered-in-space differencing, then the grid points connected by the dotted lines are completely independent from the grid points unconnected by the lines. If only the grid points connected by the dotted lines are retained, the result is a grid that is staggered-in-time.

$$
\begin{array}{ccccc}
\eta & v & \eta & v & \eta \\
{}^{\circ} & {}^{\circ}_{u} & {}^{\circ} & {}^{\circ}_{u} & {}^{\circ} \\[4pt]
u & & u & & u \\
{}^{\circ}_{v} & {}^{\circ}_{\eta} & {}^{\circ}_{v} & {}^{\circ}_{\eta} & {}^{\circ}_{v} \\[4pt]
\eta & v & \eta & v & \eta \\
{}^{\circ} & {}^{\circ}_{u} & {}^{\circ} & {}^{\circ}_{u} & {}^{\circ} \\[4pt]
u & & u & & u \\
{}^{\circ}_{v} & {}^{\circ}_{\eta} & {}^{\circ}_{v} & {}^{\circ}_{\eta} & {}^{\circ}_{v} \\[4pt]
\eta & v & \eta & v & \eta \\
{}^{\circ} & {}^{\circ}_{u} & {}^{\circ} & {}^{\circ}_{u} & {}^{\circ}
\end{array}
$$

Fig. 8.13: The 2D staggered-in-time Eliassen grid. Variables annotated above grid points are defined at even time steps, while variables annotated below grid points are defined at odd time steps. Note that if the grid at a single time level is plotted, it will appear as a D grid while the combined plot resembles an E grid.

schemes there is no advantage to using time-staggered grids. Time-staggered grids have not been widely used even for the leapfrog scheme, primarily due to the complexity of programming the finite-difference equations, and to the increased use of higher-order time-differencing schemes beyond the leapfrog scheme.

Exercises

Ex. 8.1:
a. Substitute (8.22) through (8.24) into (8.19) through (8.21) and re-arrange to obtain (8.25).
b. Find the determinant of (8.27) and solve for ω to get (8.28).

Ex. 8.2:   Differentiate (8.28) with respect to k to verify (8.31).

Ex. 8.3:   For linear waves, if the mean flow components $\bar{u}$ and $\bar{v}$ are included, the dispersion relation is the same as that without the mean flow with the exception of the terms $\bar{u}k$ and $\bar{v}l$ added. Verify this by substituting (8.22) through (8.24) into (8.16) through (8.18) and solve for the dispersion relation to obtain

$$\omega = \bar{u}k + \bar{v}l \pm \sqrt{gHK^2 + f^2}.$$

Ex. 8.4:
a. Multiply (8.61) by the arbitrary constant β and then add to (8.60) to obtain

$$\frac{\partial}{\partial t}(u + \beta\eta) + \bar{u}\frac{\partial}{\partial x}(u + \beta\eta) + \frac{\partial}{\partial x}(g\eta + \beta Hu) = 0. \qquad (8.85)$$

Then divide a factor α out of the last term in (8.85) to get

$$\frac{\partial}{\partial t}(u + \beta\eta) + \bar{u}\frac{\partial}{\partial x}(u + \beta\eta) + \alpha\frac{\partial}{\partial x}(g\eta/\alpha + \beta Hu/\alpha) = 0. \qquad (8.86)$$

b. Find values of α and β such that (8.86) has the form of the advection equation

$$\frac{\partial S}{\partial t} + c'\frac{\partial S}{\partial x} = 0 \qquad (8.87)$$

where $S = u + \beta\eta$ and $c' = \bar{u} + \alpha$.

Ex. 8.5: In several of the upcoming exercises dealing with numerical stability we will encounter the complex polynomial

$$\left(\lambda^2 - 1\right)^2 + \sigma^2 \lambda^2 = 0 \tag{8.88}$$

and will need to find the range of σ where $|\lambda| \leq 1$. Show that as long as $\sigma^2 \leq 4$, no root of (8.88) will have an amplitude greater than one, while if $\sigma^2 > 4$, at least one root will have an amplitude greater than one.

Ex. 8.6: The figure below shows a one-dimensional staggered grid where the u and v components of velocity are defined at collocated grid points.

The one-dimensional barotropic primitive equations without advection are

$$\frac{\partial u}{\partial t} = -g\frac{\partial \eta}{\partial x} + fv \tag{8.89}$$

$$\frac{\partial v}{\partial t} = -fu \tag{8.90}$$

$$\frac{\partial \eta}{\partial t} = -H\frac{\partial u}{\partial x}. \tag{8.91}$$

Using leapfrog time differencing and centered space differencing show that the stability condition is

$$\left| \sqrt{\frac{c^2 \Delta t^2}{d^2} + \frac{f^2 \Delta t^2}{4}} \right| \leq \frac{1}{2}. \tag{8.92}$$

Note: The following exercises can be challenging, with lots of room for algebraic and other mathematical errors. Intermediate steps and hints are provided in the Appendix. The following identities are helpful:

$$e^{\iota\theta} + e^{-\iota\theta} = 2\cos\theta \qquad (8.93)$$

$$e^{\iota\theta} - e^{-\iota\theta} = 2\iota\sin\theta \qquad (8.94)$$

$$\left(1 + e^{\iota\theta}\right)\left(1 + e^{-\iota\theta}\right) = 2 + 2\cos\theta \qquad (8.95)$$

$$\left(1 - e^{\iota\theta}\right)\left(1 - e^{-\iota\theta}\right) = 2 - 2\cos\theta \qquad (8.96)$$

$$\left(1 + e^{\iota\theta}\right)\left(1 - e^{-\iota\theta}\right) = 2\iota\sin\theta \qquad (8.97)$$

$$\left(1 - e^{\iota\theta}\right)\left(1 + e^{-\iota\theta}\right) = -2\iota\sin\theta \qquad (8.98)$$

Ex. 8.7:   Derive the dispersion relation (8.52) for barotropic waves on the A grid.

Ex. 8.8:   Derive the dispersion relation (8.55) for barotropic waves on the B grid.

Ex. 8.9:   Derive the dispersion relation (8.56) for barotropic waves on the C grid.

Ex. 8.10:   Derive the dispersion relation (8.57) for barotropic waves on the D grid.

Ex. 8.11:   Show that the stability condition for the leapfrog scheme applied to the A grid is $\left|c\Delta t / d\right| \le \sqrt{2}/2$.

Ex. 8.12:   Show that the stability condition for the B grid is $\left|c\Delta t / d\right| \le 1/2$.

Ex. 8.13:   Show that the stability condition for the C grid is $\left|c\Delta t / d\right| \le \sqrt{2}/4$.

The following exercises are even more challenging, primarily due to the number of terms in the equations:

Ex. 8.14: Derive the dispersion relation (8.58) for barotropic waves on the E grid. Hint: Assume solutions having the form $e^{\iota(kid'/2+ljd'/2-\omega t)}$.

Ex. 8.15: **a.** Show that the amplification factors for the 3rd-order Adams-Bashforth scheme applied to the B grid obey the two complex polynomials

$$\lambda^3 - (1 + \iota 23\sigma)\lambda^2 + \iota 16\sigma\lambda - \iota 5\sigma = 0 \qquad (8.99)$$

and

$$\lambda^3 - (1 - \iota 23\sigma)\lambda^2 - \iota 16\sigma\lambda + \iota 5\sigma = 0 \qquad (8.100)$$

where

$$\sigma^2 = \frac{c^2\Delta t^2}{18d^2}(1 - \cos kd \cos ld). \qquad (8.101)$$

b. Given that (8.99) and (8.100) will have $|\lambda| \leq 1$ for $|\sigma| \leq 0.06$, show that the stability condition is $|c\Delta t/d| \leq 0.18$.

Ex. 8.16:

a. Show that the amplification factors for the 3rd-order Adams-Bashforth scheme applied to the C grid obey the same two complex polynomials given by (8.99) and (8.100), only with

$$\sigma^2 = \frac{c^2\Delta t^2}{72d^2}(2 - \cos kd - \cos ld). \qquad (8.102)$$

b. Show that the stability condition in this case is $|c\Delta t/d| \leq 0.25$.

9.1 General Principles

Unless an atmospheric model's domain encompasses the entire globe, there will be limits to its lateral extent. These limits are known as *lateral boundaries*, and are *artificial boundaries* in the sense that they do not represent actual, physical entities. Models may also have *physical boundaries*, which coincide with real features such as the surface of the Earth or the tropopause.

Regardless of whether a boundary is physical or artificial it is important to correctly handle fluxes of momentum, heat, moisture, and mass across the boundary, both into and out of the model domain. The means of specifying or parameterizing these fluxes are referred to as *boundary conditions*. If appropriate boundary conditions are not specified at artificial boundaries, then a wave-like disturbance can reflect off of the boundary and move back into the model domain, altering the budgets of momentum, heat, and moisture within the domain. The primary physical boundary in atmospheric models is the surface of the Earth, and the handling of the boundary conditions at the surface is discussed in Chapter 10. The present chapter deals with boundary conditions at artificial boundaries.

For atmospheric modeling there are two important general principles that apply regarding boundaries:

1. The first principle is to keep the boundary as far away as possible from the region of interest in the model simulation. No matter how carefully the boundary conditions are specified they will not be perfectly representative of reality, since in reality there is no physical boundary. We want to limit the influence of the boundary on the region of interest in the simulation.

2. The second principle is to apply appropriate boundary conditions so that momentum, heat, moisture, and mass that approach the boundary will either pass through the boundary without reflecting, or will be damped to reduce any reflection back into the domain.

9.2   Lateral Boundary Conditions

In addition to preventing reflections of disturbances at the boundaries of the domain, the lateral boundary conditions also must account for the influence that disturbances and conditions outside of the model domain will have on the simulation inside the domain. It would be pointless to have a regional numerical weather prediction model whose domain covered the continental United States if disturbances in the westerlies moving in from the Pacific Ocean could not influence the simulation. This influence is achieved by using the output from a larger-domain model, such as a global model, to update conditions on the boundary of the smaller-domain model.

When implementing such a procedure the boundary conditions do not necessarily need to be updated at every single time step, but at some predefined interval that may range from a few minutes to several hours. A global model does not have to run to completion prior to starting a regional model, but the global model does have to remain ahead of the regional model. Note that the global model and the regional model are independent, but the output of the global model may influence the simulation within the regional model. The simulation within the regional model has no influence on the global model. This is referred to as one-way influence, and will be discussed in Section 9.6 in the context of nested grids.

9.3 The Radiation Boundary Condition

We now discuss the means of implementing a boundary condition to prevent the reflection of wavelike disturbances back into a model domain. This is known as the *radiation* or *open* boundary condition. We want to set up the equations such that they work like one-way check valves, allowing disturbances to move out of the domain without reflection.

9.3.1 Simple radiation boundary condition

The simplest means of achieving a radiation boundary condition is to apply the one-way wave equation operator for a positive-traveling wave,

$$\frac{\partial}{\partial t} + c\frac{\partial}{\partial x} = 0, \tag{9.1}$$

at the right-hand boundary, and at the left hand boundary the one-way wave equation operator for a negative-traveling wave,

$$\frac{\partial}{\partial t} - c\frac{\partial}{\partial x} = 0. \tag{9.2}$$

In (9.1) and (9.2) the parameter c is the wave speed. If there were also an advective flow with speed U, then we would use

$$\frac{\partial}{\partial t} + (U + c)\frac{\partial}{\partial x} = 0 \tag{9.3}$$

at the right-hand boundary, and

$$\frac{\partial}{\partial t} + (U - c)\frac{\partial}{\partial x} = 0 \tag{9.4}$$

at the left boundary.

9.3.2 Application in a 1D model

We illustrate this simple radiation boundary condition for a 1D barotropic primitive equation model on the one-dimensional staggered grid shown in Fig. 9.1. On this grid there are a total of NX grid points for η, with indices numbered from 0 through $NX - 1$.

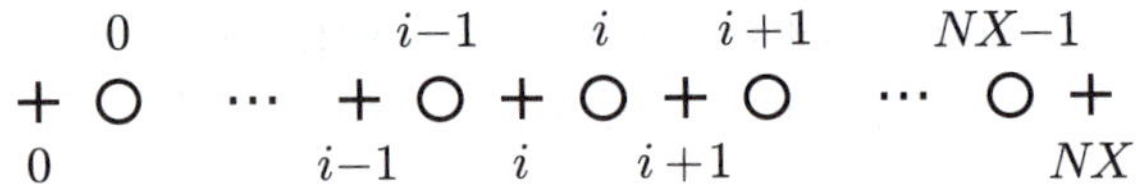

```
       0              i-1    i    i+1    NX-1
    +  O   ···    +  O  +  O  +  O   ···   O  +
       0              i-1    i    i+1          NX
```

Fig. 9.1: Staggered 1D grid with velocity grid points on the boundaries. Indices below the grid points correspond to the u grid points (denoted by '+'), while those above correspond to the η grid points (denoted by 'O'). There are a total of NX grid points for η, and $NX + 1$ grid points for u.

There is one additional grid point for u, and the u grid points are numbered from 0 through NX. The 1D primitive equations without advection or Coriolis are

$$\frac{\partial u}{\partial t} = -g\frac{\partial \eta}{\partial x} \tag{9.5}$$

$$\frac{\partial \eta}{\partial t} = -H\frac{\partial u}{\partial x}, \tag{9.6}$$

and their finite-difference forms using 3rd-order Adams-Bashforth time differencing and centered space differencing are

$$\frac{u_i^{n+1} - u_i^n}{\Delta t} = -\frac{g}{d}\left[\begin{array}{c} 23\left(\eta_{i+1}^n - \eta_i^n\right) \\ -16\left(\eta_{i+1}^{n-1} - \eta_i^{n-1}\right) \\ +5\left(\eta_{i+1}^{n-2} - \eta_i^{n-2}\right) \end{array}\right] \tag{9.7}$$

$$\frac{\eta_i^{n+1} - \eta_i^n}{\Delta t} = -\frac{H}{d}\left[\begin{array}{c} 23\left(u_i^n - u_{i-1}^n\right) \\ -16\left(u_i^{n-1} - u_{i-1}^{n-1}\right) \\ +5\left(u_i^{n-2} - u_{i-1}^{n-2}\right) \end{array}\right]. \tag{9.8}$$

Equations (9.7) and (9.8) can be applied at every grid point in the domain except for the u grid points with indices $i = 0$ and $i = NX$. If we use a *fixed* or *rigid* boundary condition such that the u values on the boundary remain zero, then when waves reach the boundary they reflect back into the domain as shown in the top and middle panels of Fig. 9.2. If we apply the one-way wave operators, (9.1) and (9.2), to the u values at $i = 0$ and $i = NX$, then for these grid

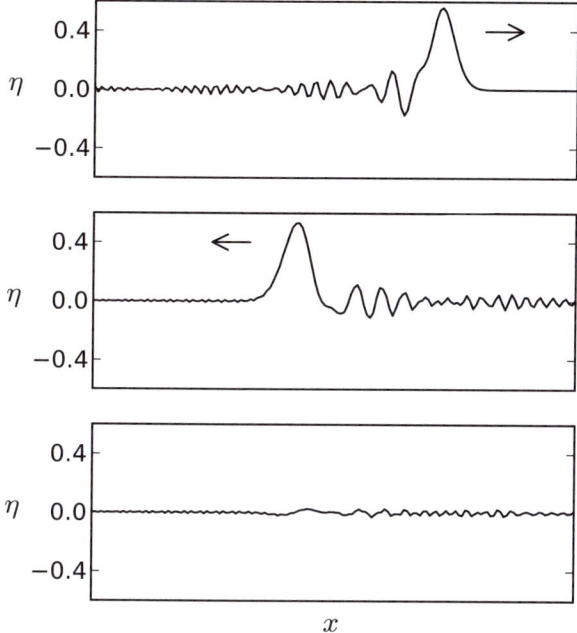

Fig. 9.2: The top panel shows a disturbance approaching the right-hand boundary of the grid. The middle panel shows the resulting reflected disturbance if rigid boundary conditions are used. The bottom panel shows the resulting reflected disturbance if the simple, one-way wave equation (9.1) is used as the boundary condition.

points the finite-difference equations are

$$\frac{u_0^{n+1} - u_0^n}{\Delta t} - c\frac{u_1^n - u_0^n}{d} = 0 \tag{9.9}$$

$$\frac{u_{NX}^{n+1} - u_{NX}^n}{\Delta t} + c\frac{u_{NX}^n - u_{NX-1}^n}{d} = 0 \tag{9.10}$$

where $c = |\sqrt{gH}|$. In this case the waves pass through the boundary with little reflection, as shown in the bottom panel of Fig. 9.2. Note that the boundary condition greatly reduces reflection but does not completely eliminate it.

In the previous example the one-way wave equation was only applied to u, and not to η. This is because the only boundary points are those for u. The first and last η grid points are essentially interior grid points, since the finite-difference equation (9.8) can be applied at these points with no modification required. If instead we had used a grid with one more η grid point than u grid points, so that the boundary grid points were η grid points, then we would apply the one-way wave equation to η, and not u.

9.3.3   Application in a 2D model

The wave speed c in (9.1) and (9.2) is actually the speed of the disturbance perpendicular to the boundary. In the 1D barotropic PE model this speed is simply

$$c = |\sqrt{gH}|. \tag{9.11}$$

In a 2D model the disturbances can approach the boundary obliquely from any angle, and so there are an infinite number of different wave speeds that must be accounted for. However, simply using (9.11) as the wave speed along the boundary in a 2D model still reduces the wave reflection, even if the Coriolis term is included. This is illustrated in Fig. 9.3, which shows a disturbance approaching the boundaries in a square 2D barotropic PE model that includes Coriolis. The model uses an Arakawa C grid with the u grid points on the East and West boundaries, and the v grid points on the North and South boundaries.

9.4   Unknown Wave Speed

In the case of a 1D barotropic PE model we have prior knowledge of what wave speed, c, to use in the one-way wave equation for the boundary conditions. Even if we include the Coriolis term and move to two dimensions, using the barotropic gravity wave speed c still reduced reflections as shown in Fig. 9.3. However, in many modeling applications we do not have prior knowledge of the wave speeds as they approach the boundary. This may be due to the model supporting many different types of waves with differing wave speeds, or supporting very dispersive waves. We therefore need a boundary condition scheme that estimates the speed of the waves as they approach the boundaries. One such scheme is the *modified Orlanski* boundary condition.[1]

---

[1]This scheme is described in detail in Tang, Y. and R. Grimshaw, 1996: Radiation boundary conditions in barotropic coastal ocean numerical models, *Computational*

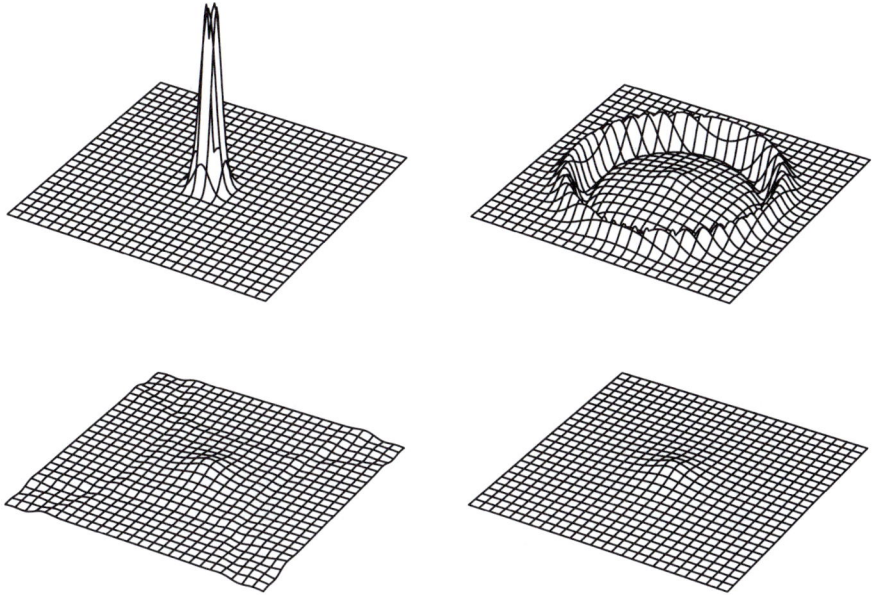

Fig. 9.3: The initial disturbance (upper-left) approaches the boundary of the 2D grid (upper-right) and reflects (lower-left). Each time the reflections interact with the boundaries, the reflections get smaller. The bottom-right panel is after much time has elapsed, so that all that remains is the steady-state solution. (Since the Coriolis effect is nonzero, the steady state consists of a slight bulge supported by a gradient-balanced anticyclonic circulation.)

In this scheme the equation for the velocity component at grid point 0 (such as at the left boundary of a 1D grid, or the West or South boundary of a 2-D grid)[2] is

$$u_0^{n+1} = u_1^n + s(u_0^n - u_1^{n+1}). \tag{9.12}$$

The parameter s in (9.12) is determined by first defining a new parameter $\hat{s}$ using either

$$\hat{s} = \frac{u_1^n - u_2^{n-1}}{u_1^{n-1} - u_2^n} \tag{9.13}$$

Physics, **123**, 96-110. This article is the primary reference for our discussion.

[2]If this were for the South boundary of a 2D grid, then we would use v in place of u, but all the other steps would be the same.

or

$$\hat{s} = \frac{u_1^{n+1} - u_2^n}{u_1^n - u_2^{n+1}}. \tag{9.14}$$

If (9.13) is used, then this is called the *explicit* modified Orlanski boundary condition, while if (9.14) is used, then it is called the *implicit* modified Orlanski boundary condition. Once $\hat{s}$ is determined, s is found as follows:

$$s = \begin{cases} \hat{s}, & \text{if } 0 < \hat{s} < 1; \\ 1, & \text{if } |\hat{s}| \geq 1; \\ 0, & \text{if } -1 < |\hat{s}| < 0. \end{cases} \tag{9.15}$$

To apply the modified Orlanski boundary condition at grid point M (the right-hand boundary of a 1D grid, or the East or North boundary of a 2D grid)[3] the equation is

$$u_M^{n+1} = u_{M-1}^n + s \left(u_M^n - u_{M-1}^{n+1} \right) \tag{9.16}$$

where the parameter s is still found from (9.15), but with $\hat{s}$ determined using either the explicit equation

$$\hat{s} = \frac{u_{M-1}^n - u_{M-2}^{n-1}}{u_{M-1}^{n-1} - u_{M-2}^n} \tag{9.17}$$

or the implicit equation

$$\hat{s} = \frac{u_{M-1}^{n+1} - u_{M-2}^n}{u_{M-1}^n - u_{M-2}^{n+1}}. \tag{9.18}$$

9.5   Absorbing (Sponge) Layers

Another way of eliminating the reflection of waves back into the model domain is to use an absorbing layer, or *sponge layer*, adjacent to the boundary, which damps any disturbances that approach the boundary. A sponge layer may be implemented in several different

---

[3]For the North boundary of a 2D grid we would use v in place of u.

ways. One method is to add an artificial viscosity term to the equations, which for the u-momentum equation would have the form

$$\frac{\partial u}{\partial t} = [\text{other terms}] + \alpha(x)\frac{\partial^2 u}{\partial x^2} , \qquad (9.19)$$

where α represents an artificial viscosity.

Another method is to include a *Rayleigh damping* term, which for the u-momentum equation would have the form

$$\frac{\partial u}{\partial t} = [\text{other terms}] - R(x)(u - u_0), \qquad (9.20)$$

where R represents the Rayleigh damping coefficient and u_0 is a background or undisturbed value of u. When implementing either (9.19) or (9.20), the values of α or R would be zero in the center of the domain, and would gradually increase as the boundaries are approached. This gradual increase is very important. If the values are increased too rapidly or abruptly, then disturbances would actually reflect off of the leading edge of the sponge layer.

Sponge layers are most commonly implemented at the top of an atmospheric model to avoid reflection of vertically propagating waves back downward into the model domain.

9.6 Variable Grid Spacing

In general, a smaller grid spacing should result in a more accurate model simulation. However, a decrease in grid spacing also results in increased data storage requirements and requires shorter time steps and more calculations to be performed, increasing the time required for computations. One strategy is to have small grid spacing in regions of interest within the domain, and coarser grid spacing elsewhere. Two common methods of achieving this are either to use a *stretched grid* or to use *nested grids*.

9.6.1 Stretched grids

A stretched grid varies the horizontal grid spacing within a single model domain so as to have the finest grid resolution in the region of interest. Stretched grid models are fairly easy to implement. Instead of using a constant grid spacing d, the grid spacing is a function of the grid point indexes and is denoted as $d_{i,j}$. Another method is to define a stretching factor at each grid point, $a_{i,j}$. The stretching factor is a number greater than one. In the finite-difference equations, whenever the grid spacing d is encountered it is then divided by the stretching factor.

Care must be exercised when using a stretched grid, to ensure that the grid spacing changes very gradually. An abrupt change in grid spacing will result in some of the energy of a wave reflecting back in the opposite direction.

Vertically-stretched grids are very commonly used in atmospheric models, with smaller grid spacing used in the boundary layer near the surface of the model, and also in the vicinity of the tropopause.

9.6.2   Nested grids

Many models now make use of high-resolution subgrids over a region of interest, embedded within a courser grid. This is referred to as *grid nesting*. Boundary conditions must be employed at the lateral boundaries of each subgrid. The boundary conditions need to not only preclude the reflection of disturbances back into the subgrid, but also allow fluxes of momentum, heat, moisture, and mass to be exchanged between the outer and inner grids. This exchange may be either *one-way* or *two-way influence*.

In one-way influence, disturbances in the outer domain can propagate into the inner domain and influence the simulation within the inner domain, but the inner domain disturbances do not propagate into the outer domain. This is similar to the way in which the output from a global model is used to update the boundary conditions in a regional model. In two-way influence, not only can disturbances in the outer domain propagate into and influence the inner domain, but inner-domain disturbances can propagate into and influence the outer domain.

If nested grids are interactive, then the ratio of the coarse-grid spacing (d_c) to fine-grid spacing (d_f) must be an odd integer[4],

$$d_c/d_f = 3, 5, 7, 9, \cdots . \tag{9.21}$$

This ensures that all fine-mesh grid corner grid points coincide with a coarse-mesh grid point so that the boundaries of the fine grid align

---

[4]This assumes that the grid spacing is uniform within each grid, and is not variable.

with a row or column of coarse-grid points. This restriction requires that the nested grid dimensions in both horizontal directions must be in the set of $3N + 1$, where N is a positive integer.

In nested-grid models, care must be taken at the boundaries to ensure the solutions in each grid are consistent with each other along the boundaries. Often, some sort of wavelength-selective sponge layer or filter is used that allows low-frequency waves to pass between the grids, but not the high-frequency waves, which are most likely to be in error and to cause reflections.

9.6.3 Adaptive grids

Another approach for concentrating high resolution over a region of interest is the *adaptive-grid* technique. In this method the model grid is not static, but changes as the model runs in response to areas of interest that may arise during the simulation. For example, if a thunderstorm develops in the simulation, then the model can automatically detect the presence of the storm and reposition the high-resolution portion of the grid.

9.7 Boundaries and Numerical Instability

Lateral boundaries are often a source of numerical instability in models. It is possible that certain boundary conditions may result in a reflected wave becoming numerically unstable and amplifying after interaction with the boundary. This often manifests itself as spurious convective activity along the boundary of an atmospheric model. This may be suppressed through application of a sponge layer or by a more careful application of the boundary conditions.

Exercises

Ex. 9.1:   Build a 1D model to solve the barotropic primitive equations without advection or Coriolis. Use the staggered grid shown in Fig. 9.1 with 401 grid points for η, and 402 grid points for u. Use a grid spacing of 3000 meters and a time interval of 30 seconds. Use a depth H of 5 meters. For the initial conditions use $u = 0$, and for

η use

$$\eta = \begin{cases} 0, & \text{for } 0 \le i \le 194; \\ 0.9, & \text{for } 195 \le i \le 205; \\ 0, & \text{for } 206 \le i \le 400. \end{cases}$$

Run the model for:

a. Fixed boundary conditions.

b. Simple radiation boundary conditions, using (9.1) and (9.2). Compare your model output to Fig. 9.2, in which the top panel is at time level $n = 3360$.

Ex. 9.2: Build a 1D model to solve the barotropic primitive equations with Coriolis, shown below:

$$\frac{\partial u}{\partial t} = -g\frac{\partial \eta}{\partial x} + fv \tag{9.22}$$

$$\frac{\partial v}{\partial t} = -fu \tag{9.23}$$

$$\frac{\partial \eta}{\partial t} = -H\frac{\partial u}{\partial x}. \tag{9.24}$$

Use a staggered grid similar to Fig. 9.1 with 401 grid points for η and 402 grid points for u and v. The u and v grid points are collocated. Use a grid spacing of 3000 meters, a time interval of 30 seconds, and the simple radiation boundary condition for u. Note that even though the v data points are on the boundary it is not necessary to apply a boundary condition to v because the same finite-difference form of (9.23) used to predict v on the interior points also works at the boundary points.

Use a depth H of 10 meters and a Coriolis parameter f of 10^{-4}s^{-1}. For the initial conditions use $u = 0$, $v = 0$, and

$$\eta = \begin{cases} 0.9, & \text{for } 0 \le i \le 199; \\ -0.9, & \text{for } 200 \le i \le 400. \end{cases}$$

The initial conditions for η, as well as the results at $n = 32,000$, are shown below.

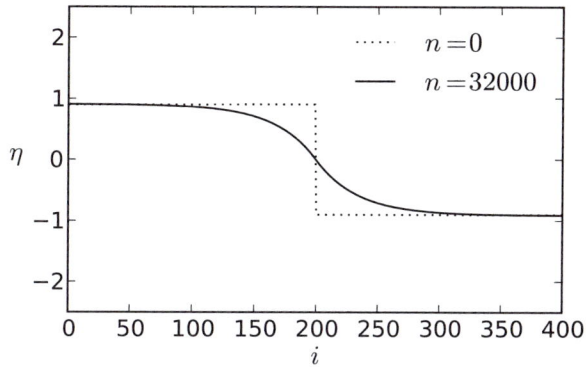

Ex. 9.3: Build a 2-D barotropic PE model on an Arakawa C grid, with the following parameters:

η grid (NX, NY):	$(101, 101)$
u grid (NX, NY):	$(101, 100)$
v grid (NX, NY):	$(100, 101)$
grid spacing (d):	3000 m
time interval (Δt):	30 s
depth (H):	10 m
Coriolis parameter (f):	$2 \times 10^{-4}\ \text{s}^{-1}$

For initial conditions use $u = 0$, $v = 0$, and

$$\eta = 0.9 \exp\left(-r^2/2\sigma^2\right),$$

where

$$r = \sqrt{(i - 50)^2 + (j - 50)^2}$$

$$\sigma = 3.$$

Use the simple radiation boundary condition at all boundaries, applying the one-way wave equation to u on the East and West boundaries, and to v on the North and South boundaries. The results at time levels $n = 0, 360, 840$, and 4000 should look similar to Fig. 9.3.

## CHAPTER 10

Subgrid-scale Processes

10.1 Introduction

Because of model resolution limitations there will always be some atmospheric processes that cannot be explicitly captured. Some of these processes are critical for accurate simulation of the atmosphere, especially for predicting sensible weather variables such as temperature and precipitation. Therefore, some of the most important components of any atmospheric model are the subgrid-scale parameterization schemes for clouds, precipitation, radiation, and exchanges of momentum, heat, and moisture fluxes with the surface of the Earth.

As model resolution becomes higher there are fewer subgrid-scale processes, since more processes can be explicitly calculated. For example, synoptic scale models must parameterize convective clouds, since an individual cloud element is smaller than the grid spacing. Cloud-resolving models use such a small grid spacing that the effects of convective clouds can be explicitly calculated. However, any processes that are important to the evolution of the atmosphere and that occur on a scale that is smaller than a grid cell must be parameterized.

All parameterization schemes must represent the subgrid-scale processes in terms of variables that are explicitly resolved by the model, with the aim of producing a bulk result that represents the impact of the subgrid-scale process on the atmosphere. For example, a model using a 20 km horizontal grid spacing cannot resolve the latent heat release of an individual convective cell. However, given the model-derived values of moisture, stability, vertical motion, and other parameters, a reasonable estimate of the average release of latent heat can be made for each grid cell. In this way the subgrid-scale latent heating processes can be accounted for without explicitly calculating the latent heat from each convective cell.

Some approaches and methods of parameterization are described in this chapter. Most of the discussion will be quite general. The goal is to acquaint the reader with the relevant concepts, leaving the details of particular implementations and schemes for more advanced treatments.

10.2   Turbulence and Reynolds Averaging

Turbulent flows contain eddies ranging over a wide variety of length scales. Turbulence can be viewed as a spectrum of eddies superimposed on the mean flow, and must be represented correctly because it is a crucial mechanism for the transfer and dissipation of kinetic energy, as well as for the transport of humidity and thermal energy.

The effects of turbulence are included in the governing equations through the process of *Reynolds averaging*, whereby all of the dependent variables are written as the sum of a time-averaged mean and a perturbation around the mean. For example, the horizontal momentum and temperature variables would be written as $u = \bar{u} + u'$, $v = \bar{v} + v'$, and $T = \bar{T} + T'$. The time-averaged mean values are denoted by an overbar, while the perturbations are denoted using the prime symbol. The time-averaged mean values are themselves functions of x, y, z, and t. Once all the dependent variables are written in terms of the mean and perturbations values, they are then put into the governing equations and all terms expanded out. For example, the u-momentum equation in Cartesian coordinates becomes

$$\frac{\partial}{\partial t}(\overline{u} + u') + (\overline{u} + u')\frac{\partial}{\partial x}(\overline{u} + u') + (\overline{v} + v')\frac{\partial}{\partial y}(\overline{u} + u')$$

$$+ (\overline{w} + w')\frac{\partial}{\partial z}(\overline{u} + u') = -\frac{1}{(\overline{\rho} + \rho')}\frac{\partial}{\partial x}(\overline{p} + p')$$

$$- 2\Omega \sin\phi(\overline{u} + u') - -2\Omega \cos\phi(\overline{w} + w'). \quad (10.1)$$

Once this substitution has been made, the resulting equation is again time-averaged, with the following rules applied:

1. The time average of a term containing only time-averaged quantities will just be equal to the original term (the average of an average is just the average.) For example:

$$\overline{2\Omega \sin\phi\overline{u}} = 2\Omega \sin\phi\overline{u}$$

$$\overline{\overline{v}\frac{\partial\overline{u}}{\partial x}} = \overline{v}\frac{\partial\overline{u}}{\partial x}.$$

2. The time average of a term containing only a single perturbation quantity will be zero (the perturbations are assumed to have a zero mean.) For example:

$$\overline{2\Omega \sin\phi u'} = 0$$

$$\overline{\overline{v}\frac{\partial u'}{\partial x}} = 0.$$

3. The quantity $\overline{\rho} + \rho'$ can be approximated simply as $\overline{\rho}$, unless it appears in a term involving buoyancy. This assumption is very similar to the Boussinesq approximation.

Applying these rules allows several of the terms in the equation to disappear, and the resulting u-momentum equation[1] is

$$\frac{\partial \overline{u}}{\partial t} + \overline{u}\frac{\partial \overline{u}}{\partial x} + \overline{v}\frac{\partial \overline{u}}{\partial y} + \overline{w}\frac{\partial \overline{u}}{\partial z} =$$

$$- \frac{1}{\rho}\frac{\partial \overline{p}}{\partial x} - 2\Omega \sin\phi\,\overline{u} - 2\Omega \cos\phi\,\overline{w}$$

$$- \left[\frac{\partial \overline{(u'u')}}{\partial x} + \frac{\partial \overline{(v'u')}}{\partial y} + \frac{\partial \overline{(w'u')}}{\partial z}\right]. \quad (10.2)$$

Equation (10.2) looks nearly identical to the non-Reynolds averaged u-momentum equation, with the exception that the time-averaged variables appear in place of the original variables, and there are three additional term (in brackets) that involve time averages of products of the perturbations. These bracketed terms are associated with turbulent fluxes of momentum, and are fully discussed in Section 10.2.2.

10.2.1   Weather observations

When a weather observation is taken it is not the instantaneous value of wind speed that is recorded, but instead an average speed taken over several minutes.[2] In nearly every meteorology course and textbook, including the present text, it is the time-averaged (Reynolds-averaged) values of u, v, and w that are assumed to be in use, even though it is not explicitly stated as such. The turbulence terms are not usually included because they have negligible contributions in the free troposphere except in regions of very high wind shear.

10.2.2   Turbulent fluxes

The bracketed terms appearing in (10.2) represent the transport of momentum by turbulent eddies, and can properly be referred to as the *turbulent momentum flux* terms. In the Reynolds-averaged thermodynamic energy equation we encounter terms involving *turbulent heat fluxes*, of the form $\overline{u'T'}$, $\overline{v'T'}$, and $\overline{w'T'}$. Likewise, in the

---

[1]At the scale of turbulent eddies it is appropriate to use the incompressible continuity equation, and with such it can be shown that $\overline{u'\frac{\partial u'}{\partial x}} + \overline{v'\frac{\partial u'}{\partial y}} + \overline{w'\frac{\partial u'}{\partial z}} = \frac{\partial \overline{(u'u')}}{\partial x} + \frac{\partial \overline{(v'u')}}{\partial y} + \frac{\partial \overline{(w'u')}}{\partial z}$.

[2]In the U.S. a standard synoptic wind observation is taken as a two-minute mean wind speed.

water-mass continuity equation there are terms involving *turbulent moisture fluxes*, $\overline{u'q'}$, $\overline{v'q'}$, and $\overline{w'q'}$.

Depending on the physical circumstances of an atmospheric flow the turbulent fluxes may be *isotropic*, meaning that they have the same order of magnitude in all directions, or they may be *anisotropic* (larger in one direction). It is very typical, particularly near the surface of the Earth, for the turbulent fluxes to be much greater in magnitude vertically than horizontally. In these circumstances the terms involving vertical derivatives such as $\partial\overline{(w'u')}/\partial z$, $\partial\overline{(w'T')}/\partial z$, and $\partial\overline{(w'q')}/\partial z$ will be more important than those involving horizontal derivatives.

10.2.3 Closure

The turbulent flux terms appearing in the governing equations present a problem because they introduce additional dependent variables involving perturbation quantities. The introduction of these additional variables then requires additional equations in order to form a closed system. These additional equations must somehow relate the perturbation variables to the time-averaged variables. There is no single perfect method or set of equations to accomplish this relation. The particular set of equations used to relate the perturbation variables to the time-averaged variables is known as the *closure* scheme.

One example of a closure scheme would be to parameterize the fluxes in terms of the gradients of the time-averaged quantities, using equations of the form

$$\overline{w'u'} = -K_u\frac{\partial\overline{u}}{\partial z}; \quad \overline{w'v'} = -K_u\frac{\partial\overline{v}}{\partial z};$$
$$\overline{w'T'} = -K_T\frac{\partial\overline{T}}{\partial z}; \quad \overline{w'q'} = -K_q\frac{\partial\overline{q}}{\partial z}. \tag{10.3}$$

This closes the system of equations by introducing four new equations to balance the four unknown eddy fluxes.[3] The parameter K_u is the *eddy viscosity*, and K_T, and K_q are the *eddy diffusivities* of heat,

---

[3]At first glance it may seem that we need five equations, since it appears we have five new unknowns, u', v', w', T', and q'. However, the perturbations never appear singly, and we can consider the four covariances of the perturbations, $\overline{w'u'}$, $\overline{w'v'}$, $\overline{w'T'}$, and $\overline{w'q'}$, to be the new unknowns.

and moisture. These parameters may be taken to be constants, or as functions of location in the model domain. They may themselves even be parameterized in terms of other model variables.

Equations (10.3) are known as a *local closure* scheme, because they involve only information near the locality of the model grid point, and not information elsewhere in the domain. If a scheme requires information away from the local grid point, then it is called a *non-local closure* scheme. Equations (10.3) are also an example of a *first-order* closure scheme, because they use only *first-order moments* (or *covariances*) of the perturbation variables. In higher-order closure schemes the first-order moments may be written in terms of higher-order moments (terms involving averages of three or more perturbation values) such as $\overline{u'u'w'}$ or $\overline{u'w'T'}$, which then are parameterized in terms of the model variables. If a particular closure scheme requires that these second-moments be parameterized, it is called a *second-order* closure scheme. Some schemes require a combination of first-order and second-order moments to be parameterized in terms of the time-averaged variables, in which case they are called *one-and-a-half-order* closure schemes.

10.3 Clouds and Convection

10.3.1 General considerations

Convection and clouds alter the vertical temperature profile of the atmosphere through latent heat release and transport of thermal energy. Convection also redistributes moisture. The two equations where these processes appear are the thermodynamic energy equation,

$$\frac{\partial T}{\partial t} + \vec{V} \cdot \nabla T - \omega/c_p\rho = \dot{Q}_R + \dot{Q}_L + \dot{Q}_T, \qquad (10.4)$$

and the water-mass continuity equation,

$$\frac{\partial(\rho q)}{\partial t} + \nabla \cdot (\rho q \vec{V}) = S_O - S_K + S_T. \qquad (10.5)$$

In (10.4) $\dot{Q}_R$ represents radiative heating or cooling, $\dot{Q}_L$ represents latent heating or cooling, and $\dot{Q}_T$ represents the effects of turbulent heat fluxes. In (10.5) S_O represents sources of water vapor into the air through evaporation or sublimation; S_K represents sinks through condensation, deposition, or precipitation; and S_T represents the effects of turbulent moisture fluxes. The terms $\dot{Q}_L$, S_O, and S_K are not independent, since many of the processes through which atmospheric water vapor is gained or lost are also associated with latent heat release or absorption.

All of the terms on the right-hand side of (10.4) and (10.5) are subgrid-scale, and must be parameterized. The $\dot{Q}_R$ term will be handled by the radiation parameterization scheme, while the $\dot{Q}_T$ and S_T terms will be taken care of by the turbulence closure and boundary layer parameterization schemes. The $\dot{Q}_L$ term is handled by the cloud and convection parameterizations schemes. The terms S_O and S_K are handled jointly by the cloud and convection schemes and the boundary layer scheme.

The latent heat released during convection depends on several parameters such as updraft velocity, vertical depth and duration of convection, the horizontal scale of the convection, and the temperature and saturation ratio of the air. The static stability of the temperature profile, convective available potential energy (CAPE), and convective inhibition (CIN) are also important parameters. An effective parameterization scheme must account for all of these parameters, as well as adjust for factors which will reduce updraft velocities, such as aerodynamic drag, entrainment of drier air into the convection, evaporative cooling, liquid water load, and compensating subsidence. The end result of the convective parameterization scheme is to come up with a modified temperature and humidity profile that is valid within the model grid cell. More sophisticated convective parameterization schemes not only parameterize the effects of latent heating and moisture redistribution, but include detailed cloud microphysics parameterizations for the different forms of liquid and solid cloud and precipitation types that can occur within convective clouds.

10.3.2   Types of convective parameterizations

There are many convective parameterization schemes available for use in atmospheric models. The choice of optimal scheme depends on the scale of the model, as well as on the purpose of the model simulation. A scheme that works well for high-desert, summer monsoon convection may not be suited for a model that is used for tropical cyclones.

Many convective clouds, such as stratocumulus and cumulus humilis, are shallow and nonprecipitating, and are adequately parameterized using a *shallow convection scheme*. These schemes often assume uniform clouds with strong entrainment and no downdrafts. Important parameters are the rate of destabilization due to heating and cooling, and also the fraction of the grid cell that contains cloud.

In large-scale models, the latent heat released by convection is expected to be controlled by the horizontal convergence of moisture. The amount of heating may therefore be parameterized in terms of the large-scale horizontal moisture convergence. The vertical placement of the heating is determined by the model's vertical moisture profile. The *Anthes-Kuo* schemes are examples of this type of convective parameterization.

Another method for large-scale models is that of *convective adjustment*, in which idealized, post-convection temperature and humidity profiles are assumed, and the model's temperature and humidity profiles are nudged toward these ideal profiles. Different idealized profiles can be used depending on whether the convection is shallow or deep, and also depending on season and geographic location. The *Betts-Miller* scheme is of this type.

Convective heat release in the upper atmosphere serves to stabilize the atmosphere. Some schemes assume that there is an equilibrium between this convective stabilization and the destabilization due to large-scale forcing. These schemes use this equilibrium to calculate the amount of latent heat released by the convection. The distribution of the latent heating in the model must also be parameterized, and is often based on the updraft and downdraft fluxes of moisture and heat within the convection. The *Grell* and *Arakawa-Schubert* schemes are both of this type. The main differences between the two are that the Grell scheme uses the fluxes from a single cloud, whereas the latter bases the fluxes on a population of clouds.

Another method for determining the amount of convective heating is to calculate the magnitude and duration of convection necessary to remove a specified fraction of the CAPE from the model sounding. This methodology was developed and is used in the *Kain-Fritch-Chappel* family of schemes.

10.4 Radiation

Radiative processes have direct and indirect impacts on the dynamical, physical, and chemical processes in the atmosphere.[4] Topography significantly complicates parameterization of the radiation fluxes at the Earth's surface, since irregular terrain results in varying slope aspect, slope angle, sky view factor and shadowing, all of which alter the radiation field. Similar complexities are created by irregular cloud fields. As spatial resolutions of models increase, these effects gain importance. Radiation parameterization schemes are designed to estimate the molecular scale radiative processes and topographic influences in order to determine the heating and cooling rates of the modeled atmosphere. Radiation parameterizations are also important for models that include chemical reactions, particularly those that are driven by photolysis from sunlight.

10.4.1 Radiative transfer equation

The fate of a beam of radiation passing through the atmosphere is governed by the radiative transfer equation,

$$dI_\lambda = -\beta_\lambda I_\lambda ds + \beta_\lambda J_\lambda ds, \tag{10.6}$$

where I_λ is the monochromatic intensity, β_λ is the volume extinction coefficient, s is the coordinate of length measured along the path, and J_λ is the *source function*. The first term on the right-hand side of (10.6) represents a loss of radiation from the beam due to absorption, as well as the scattering of photons out of the beam. The second term on the right-hand side represents a gain of radiation into the beam due to either emission of radiation by the atmosphere along the path, or due to scattering of radiation into the beam.

The volume extinction coefficient can be written in terms of either: (1) the number density, n, of the molecules responsible for the extinction and the extinction cross section of the molecules, σ_λ, or (2) the mass density, ρ, and the mass-extinction cross section of the molecules, k_λ. These relations are given by

$$\beta_\lambda = n\sigma_\lambda = \rho k_\lambda. \tag{10.7}$$

---

[4]A more comprehensive treatment may be found in the companion volume by Petty, G., 2006: *A First Course in Atmospheric Radiation*, Sundog Publishing, 452 pp.

The extinction coefficients β_λ, σ_λ, and k_λ apply to specific types of molecules (oxygen, nitrogen, water vapor, etc.). However, for a mixture of gases an effective extinction coefficient for the mixture can be found through suitable weighted averages, and so can be applied to either a homogeneous gas or a mixture of gases.

10.4.2   The plane-parallel atmosphere

Models often use the assumption that for radiative transfer purposes the atmosphere is horizontally homogenous. This is called the *plane-parallel* assumption, and allows the radiative transfer equation to be written in terms of altitude, z, and the cosine of the zenith angle,

$$\mu = \cos \theta. \tag{10.8}$$

For upward propagating radiation

$$ds = dz/\mu, \tag{10.9}$$

while for downward propagating radiation

$$ds = -dz/\mu. \tag{10.10}$$

The radiative transfer equation in a plane-parallel atmosphere then becomes

$$dI_\lambda^+ = -\beta_\lambda I_\lambda^+ dz/\mu + \beta_\lambda J_\lambda^+ dz/\mu \tag{10.11}$$

for upward-propagating radiation, and

$$dI_\lambda^- = \beta_\lambda I_\lambda^- dz\mu - \beta_\lambda J_\lambda^- dz/\mu \tag{10.12}$$

for downward propagation.

10.4.3   Shortwave versus longwave radiation

Atmospheric radiation is conveniently divided into two basic types based on whether its source is emission from the Sun, or emission from the Earth and its atmosphere. Solar radiation is referred to as *shortwave* radiation, and has wavelengths primarily in the ultraviolet, visible, and near-infrared. Terrestrial radiation is called *longwave* radiation, and is primarily emitted in the far-infrared and microwave portion of the spectrum. The wavelength cutoff between short- and longwave radiation is usually placed at four microns.

The radiative transfer properties of short and longwave radiation are quite different, and the radiative transfer equation can be tailored to each type of radiation. For longwave radiation the effects of scattering can be neglected, and the source function, J_λ, can often be approximated using Planck's Law for monochromatic blackbody radiation. For shortwave radiation, emission by the atmosphere can usually be neglected in the source function.

10.4.4   Optical thickness and optical depth

We can ignore emission of shortwave radiation by the atmosphere. If we also ignore multiple scattering, so that no photons can be scattered back into the beam, the source function completely disappears from the radiative transfer equation, which then becomes

$$dI_\lambda = -\beta_\lambda I_\lambda\, ds. \tag{10.13}$$

Integrating along the path from s_0 to s_1 yields

$$I_\lambda(s_1) = I_\lambda(s_0) \exp\left(-\int_{s_0}^{s_1} \beta_\lambda\, ds\right). \tag{10.14}$$

The dimensionless argument for the exponential in (10.14) is called the *path optical thickness*, τ_s,

$$\tau_s = \int_{s_0}^{s_1} \beta_\lambda\, ds. \tag{10.15}$$

The *vertical optical thickness*, τ_z, is the path optical thickness for a vertical path,[5]

$$\tau_z = \int_{z_0}^{z_1} \beta_\lambda\, dz. \tag{10.16}$$

In a plane-parallel atmosphere, optical thickness and vertical optical thickness are related via

$$\tau_s = \tau_z/\mu. \tag{10.17}$$

The vertical optical thickness between some altitude z and the top of the atmosphere is called the *optical depth*, τ_d, and is given by

$$\tau_d(z) = \int_z^\infty \beta_\lambda\, dz. \tag{10.18}$$

---

[5]Optical thickness must be a positive value, implying that z_1 must be greater than z_0 in this definition.

Optical thickness is a measure of the transparency of the atmosphere to passage of radiation. Contrary to the implications of the name of optical 'thickness', it is a dimensionless number and not a length. Optical thickness increases with path length, but it can also increase due to clouds and aerosols, without increasing the path length.

10.4.5   Radiative flux

So far we have been dealing with *monochromatic* radiation, having a specific wavelength. By integrating over all wavelengths we have the *total* intensity,

$$I = \int_0^\infty I_\lambda \, d\lambda. \tag{10.19}$$

While intensity is a measure of the radiative power in a beam coming from a certain direction, flux is the radiative power incident on a horizontal unit of area. The flux includes contributions of radiative power coming in from all different angles over an entire hemisphere, so the flux is found by integrating over azimuth angles ranging from zero to 2π and zenith angles from zero to $\pi/2$ for downward propagation (zero to $-\pi/2$ for upward propagation),

$$F = \int_0^{2\pi} \int_0^{\pi/2} I \cos\theta \sin\theta \, d\theta \, d\phi. \tag{10.20}$$

The radiative fluxes into and out of an atmospheric layer control the heating or cooling rate of the layer. Thus, the goal of a radiation parameterization is to parameterize the radiative fluxes in terms of the other, explicitly calculated model variables.

10.4.6   General principles

Cloud cover has a large influence on the transmission of radiation through the atmosphere, and so an accurate representation of clouds is crucial for any radiation parameterization scheme to be effective. Aerosols (haze, dust, smoke particles suspended in the air) also have a significant impact on optical depth. Humidity has an impact directly on the transmission of shortwave radiation through absorption of radiation by water molecules. Humidity also indirectly impacts radiative transfer through the creation of haze particles and cloud droplets, which scatter solar radiation. In order to accurately represent the radiative transfer process a scheme must account for all of the complex radiative transfer processes that influence the amount of radiation being absorbed by the clouds, particulates and atmosphere and the amount of radiation reaching the ground.

In addition to accounting for the extinction of radiation as it passes through the atmosphere, a radiation scheme must also account for the effects of the backscattering of radiation from interactions with topography, without degrading the model's computational performance. Some schemes directly compute mean fluxes for each model grid cell based on flux computations at full spatial resolution of a digital elevation model covering the model domain. This type of scheme has the advantage in that it does not require a difficult computation of the averaged topographic properties such as aspect angles. Another advantage for this type of scheme is that it uses a nonlocal computation of sky view restriction and shadowing effects.

10.4.7 Actinic flux

For most modeling applications it is the heating and cooling rates of an atmospheric layer that are the results desired of a radiation parameterization scheme. However, many chemical reactions in the atmosphere are *photolytic*, meaning that they require the absorption of solar radiation for splitting molecular bonds . These reactions are sensitive to the amount of solar radiation at specific wavelengths, and to adequately model these reactions the solar flux needs to be known. There is a subtle, but important, distinction that must be made when calculating solar fluxes for such chemical reactions. It is not the net flux into/out of a horizontal layer of air that is important, but instead the *actinic flux*, which is the flux into a complete sphere surrounding a point in space.[6]

10.5 Planetary Boundary Layer

Although the sun is the direct energy source for driving atmospheric motions, most of this energy is not directly absorbed by the atmosphere, but is instead absorbed by the Earth's surface and

---

[6]For details concerning this distinction see Madronich, S., 1987: Photodissociation in the Atmosphere: 1. Actinic flux and the effects of ground reflection and clouds, *J. Geophys. Res.*, **92**, 9740–9752.

then transferred back to the atmosphere. The surface of the Earth not only radiates and conducts thermal energy to the atmosphere, but also is the main source of water vapor (and therefore latent heat) and is a momentum sink. These transfer processes all occur within the planetary boundary layer (PBL), which is that region of the atmosphere that is directly influenced by surface friction and the turbulent exchange of the fluxes of momentum, heat and moisture. Correctly parameterizing these interactions between the atmosphere and the Earth's surface is a critical component of an atmospheric model.

The depth of the PBL is variable. It can be as shallow as 100 meters at night over land, and grow to a depth of a few thousand meters when the atmosphere is unstable and strongly heated from the surface. A PBL parameterization scheme must account for variations in momentum, heat, and moisture fluxes over widely changing conditions. Important parameters are the type of ground cover (rock, bare soil, and vegetation), topography, soil moisture, ice cover, and others. It must also account for air-sea interactions not only over oceans but over large lakes as well.

It is essential for models at all scales to realistically represent the key processes of the PBL. The budgets of momentum, heat, and moisture are controlled by surface fluxes of these parameters, on time scales ranging from hours to days. Many other subgrid-scale processes are dependent on these surface fluxes. It is therefore essential that models have a realistic representation of boundary layer processes. Also, analyzed and forecasted fields of surface fluxes are needed as input for other applications driven by the output from atmospheric models. Such applications include wave models, air pollution models and climate models.

10.5.1   Turbulent kinetic energy

An important concept for characterizing and parameterizing turbulence within the PBL is *turbulent kinetic energy (TKE)*, which is defined as the time-averaged kinetic energy per unit mass contained within the turbulent perturbations. It is denoted as e, and represented mathematically as

$$e = \frac{1}{2}\left(\overline{u'^2} + \overline{v'^2} + \overline{w'^2}\right). \tag{10.21}$$

Turbulent kinetic energy is created through either mechanical production of turbulence in sheared flow, or buoyant production due to static instability. It is generated primarily in the form of larger turbulent eddies, and the energy is transferred to smaller scales via the *energy cascade*, which is aptly summarized by L. F. Richardson's famous poem,

> Big whorls have little whorls,
> Which feed on their velocity,
> And little whorls have lesser whorls,
> And so on to viscosity.

A prognostic equation for TKE is given symbolically as

$$\frac{\overline{D}e}{Dt} = MP + BP + TR - \varepsilon, \tag{10.22}$$

in which the terms MP, BP, TR, and ε represent mechanical production, buoyant production, transport and redistribution, and dissipation respectively.

Mechanical production. The mechanical production of turbulence occurs due to dynamical instabilities in a sheared flow. This term has the form

$$MP = -\overline{u'w'}\frac{\partial \overline{u}}{\partial z} \tag{10.23}$$

and can be either positive or negative.

Buoyant production. Buoyant production represents the production of turbulence due to thermals caused by surface heating. It can also be called the *buoyancy flux*, and has the form

$$BP = \overline{w'b'}, \tag{10.24}$$

where the *buoyancy* is defined as

$$b' = -g\left(\frac{\rho'}{\rho_0}\right). \tag{10.25}$$

Like mechanical production, buoyancy production is usually largest near the surface. Under stable conditions this term will be negative and represent a loss of *TKE*.

Transport and redistribution. This term has two components, and is represented mathematically as

$$TR = -\left(\overline{u'\frac{\partial e}{\partial x}} + \overline{w'\frac{\partial e}{\partial z}}\right) - \frac{1}{\rho_0}\left(\overline{u'\frac{\partial p'}{\partial x}} + \overline{w'\frac{\partial p'}{\partial z}}\right). \qquad (10.26)$$

The first component represents advection of *TKE* by the turbulent eddies themselves. The second component represents the effects of pressure perturbations within the turbulent eddies.

Dissipation. The dissipation term has the mathematical form

$$\varepsilon = \nu\overline{\left[\left(\frac{\partial}{\partial x} + \frac{\partial}{\partial z}\right)(u' + w')\right]^2}. \qquad (10.27)$$

This term always represents a loss of turbulent kinetic energy energy due to molecular friction (ν is the kinematic viscosity).

10.5.2 Structure

The PBL itself is composed of several different sublayers, each of which has its own characteristics. Regardless of other factors, all planetary boundary layers have, at their lowest level, a very thin (order of millimeters) layer right next to the surface where molecular and viscous processes dominate. This layer is called the *viscous sublayer*. All exchanges of momentum, heat, and moisture must pass through the viscous sublayer.

Above the viscous sublayer the remaining structure of the PBL will depend on factors such as time of day, amount of surface heating, etc. Typically a daytime PBL with strong surface heating will tend toward an adiabatic lapse rate and be statically unstable due to surface heating. The daytime PBL can be divided into three layers: (1) a surface layer, (2) a convective mixed layer, and (3) an entrainment zone. In contrast, the nighttime PBL exhibits a strong static stability especially in the surface layer exhibiting a subadiabatic lapse rate and high static stability. The nighttime PBL typically exhibits only two layers: (1) a very stable surface layer and (2) a residual layer from the daytime convectively mixed layer where the lapse rate goes adiabatic near the top of the layer, but remains statically stable because of surface cooling. Figure 10.1 gives a depiction of the of the various PBL layers and conditions that must be represented in a model.

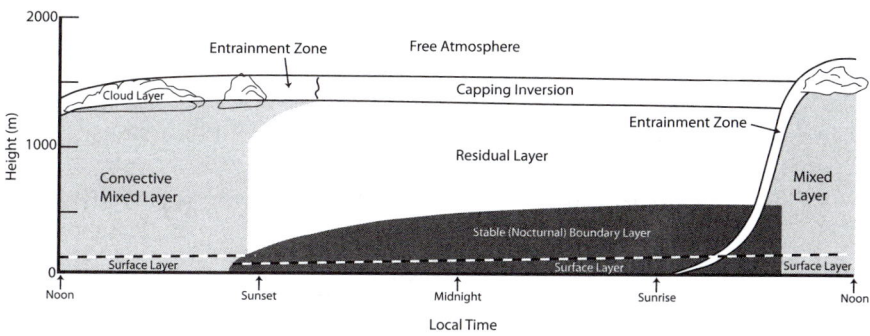

Fig. 10.1: Representation of the various PBL layers and PBL conditions depicting the evolution of the PBL during the course of the day under fair weather conditions. After Stull, R., 1988: *An Introduction to Boundary Layer Meteorology*, Springer, 670 pp.

10.5.3 Similarity theory

Boundary-layer processes are often too complicated to formulate descriptions and parameterizations based on first principles. *Similarity theory* is therefore employed. Similarity theory is based on the hypothesis that functional relationships exist among the various nondimensional groups that can be formed from the physical parameters describing a system. The functions themselves must be determined empirically. Similarity theory provides the framework to organize and group the experimental data.

To apply similarity theory we first identify the relevant physical parameters describing the system or process. Then, dimensionless groups are formed from these parameters. Experimental data is then used to find functional relations between the dimensionless groups. The physical parameters describing the PBL may be different in the lower boundary layer versus the upper boundary layer, and also may differ depending on the static stability. Different schemes are therefore usually used for different parts and types of boundary layers.

10.5.4 The neutrally-stable surface layer

In the surface layer the turbulent fluxes dominate all other terms in the equations of motion. Some assumptions made regarding the surface layer are:

- The turbulent eddies are isotropic, so that u and w are of the same order of magnitude.

- The turbulent fluxes are nearly constant with height.

- The magnitude of the turbulent fluxes can be represented by the *friction velocity, u^**, defined as

$$u^* = \left|\overline{u'w'}\right|^{1/2}. \tag{10.28}$$

- The size of the eddies is proportional to the height above the ground.

Using these assumptions the relevant factors for describing the turbulent fluxes in the neutrally-stable surface layer are altitude (z), vertical shear of the mean wind ($d\overline{u}/dz$), and the friction velocity, u^*. These three parameters can be arranged into a nondimensional group that is then set equal to a nondimensional constant, C,

$$\frac{z}{u^*}\frac{d\overline{u}}{dz} = C. \tag{10.29}$$

Integrating (10.29) upward with respect to height from level z_0, where the mean wind is zero, results in the logarithmic wind profile characteristic of the neutral surface layer,[7]

$$\overline{u}(z) = Cu^* \ln\left(z/z_0\right). \tag{10.30}$$

The altitude z_0 is called the *roughness length*, because it is related to the roughness of the surface over which the wind is blowing. It is expected that the roughness length will be smaller than, but proportional to, the mean height of the surface elements. The roughness length over a field of corn would be larger than the roughness length over a lawn. The roughness length may also change with time and be a function of u^*. For example, over water the roughness length is tied to the wave height, which is dependent on wind speed.

---

[7]The constant C is usually written as $1/\kappa$, where κ is the von Karman constant.

10.5.5   The non-neutral surface layer

In a non-neutral surface layer an additional parameter is needed to describe the static stability. This is provided by the *Monin-Obukov* length, L, which is defined as

$$L = -\frac{u*^3}{\kappa B_0} \qquad (10.31)$$

where B_0 is the surface *buoyancy flux*,

$$B_0 = \left(\overline{w'b'}\right)_0. \qquad (10.32)$$

The buoyancy flux accounts for the effects of both heat and moisture on the buoyancy of an air parcel. If the buoyancy flux is positive it implies that there is heat and/or moisture being transported upward from the surface, causing the atmosphere to be less stable. A negative buoyancy flux implies that the atmosphere is losing heat and/or water vapor to the surface, which would lead to a stabilization of the atmosphere.

The Monin-Obukov length is physically interpreted as the height above the surface where buoyant production of turbulence first dominates over the mechanical production of turbulence. For a neutrally-stable surface layer the buoyancy flux is zero, and the Monin-Obukov length is infinite. This makes sense in that for the neutral case buoyancy never dominates the production of turbulence, and so as we go upward trying to find an altitude where buoyancy dominates we would never stop. For a stable surface layer the Monin-Obukov length is positive and approaches zero as stability increases. For an unstable surface layer the length is negative and approaches zero as stability decreases.[8]

A dimensionless function ϕ, called the stability function, is postulated as being a function of the ratio of z/L. This function is included among the other parameters, z, $d\overline{u}/dz$, u^*, so that the analog of (10.29) for the non-neutral surface layer is

$$\frac{z}{u^*}\frac{d\overline{u}}{dz} = C\phi. \qquad (10.33)$$

---

[8]The Monin-Obukov length is a discontinuous function. As an unstable surface layer approaches neutrality, L approaches $-\infty$; but as a stable surface layer approaches neutrality, L approaches $+\infty$.

The form of the stability function is found empirically for various levels of stability. It is frequently given as

$$\phi(z/L) = \begin{cases} 1 + \beta z/L & \text{stable;} \\ 1 & \text{neutral;} \\ (1 - \alpha z/L)^{-1/4} & \text{unstable.} \end{cases} \qquad (10.34)$$

The constants α and β are empirically determined.

10.5.6 Upper boundary layer

Above the surface layer the fluxes are no longer constant with height, but typically decrease monotonically with altitude. This region is referred to as the upper boundary layer. The parameters important in the surface layer remain important in the upper boundary layer, but there are also additional important parameters. These are (1) the boundary layer height, h, which provides a length scale, and (2) the Coriolis parameter,[9] the inverse of which provides a time scale. Parameterization of the upper boundary layer depends on static stability and depth, with different parameterizations for neutral, convective, and stable conditions.

10.6 Gravity Wave Drag

Stably-stratified flow over topography can generate vertically propagating gravity waves, which transport momentum and energy upward to regions where they are dissipated as heat through viscous processes. This redistribution of momentum and energy is known as *gravity wave drag*, and is an important subgrid-scale process that cannot be fully resolved by models. Numerous studies have demonstrated that the addition of gravity wave drag into the formulation of a model reduces forecast biases in winds and temperature.

---

[9]The Coriolis parameter is not important for all applications. Its importance depends on latitude and the time and spatial scale of the model simulation.

## CHAPTER 11

Spectral Models

11.1 Orthogonal Basis Functions and Transforms

Spectral modeling uses orthogonal basis functions to transform partial differential equations in space and time into ordinary differential equations in time only. These ordinary differential equations are then integrated forward in time using finite differencing. In order to understand spectral modeling we need to understand orthogonal basis functions and forward and inverse transforms of functions.

11.1.1 Orthogonal basis functions

Two functions $g(x)$ and $h(x)$ are orthogonal over an interval (a, b) if the following condition holds,

$$\int_a^b gh^* dx = 0, \tag{11.1}$$

where h^* indicates the complex conjugate of the function h. The *Galerkin* method decomposes a function, $f(x, t)$, that is periodic on the interval (a, b) into a sum of orthogonal basis functions, $\varphi_l(x)$, such that

$$f(x, t) = \sum_{l=0}^{\infty} F_l(t) \varphi_l(x). \tag{11.2}$$

In (11.2) the coefficients $F_l(t)$ represent the time-dependent amplitudes for each basis function and are referred to as the *spectral coefficients*. Both the basis functions and the coefficients may be complex-valued. The coefficients are determined by multiplying (11.2) by the complex conjugate of the basis function and integrating over the interval (a, b),

$$\int_a^b f \varphi_m^* dx = \sum_{l=0}^{\infty} F_l \int_a^b \varphi_l \varphi_m^* dx. \qquad (11.3)$$

Since the basis functions are orthogonal, the only non-zero term on the right-hand side of (11.3) will correspond to $l = m$. Denoting the integral in this term as

$$\int_a^b \varphi_m \varphi_m^* dx = B_{mm}, \qquad (11.4)$$

(11.3) then becomes

$$\int_a^b f \varphi_m^* dx = F_m B_{mm} \qquad (11.5)$$

or

$$F_m = B_{mm}^{-1} \int_a^b f \varphi_m^* dx. \qquad (11.6)$$

11.1.2 Forward and inverse transforms

Equation (11.6) is called the *forward transform* of the function f. The forward transform converts the function into the spectral coefficients F_m. Equation (11.2) is referred to as the *inverse* or *backward transform* and converts the spectral coefficients back to the original function. Depending on the actual basis functions used, the transform may be given a specific name. For example, if complex exponentials are used as the basis functions, then (11.2) and (11.6) describe a form of the Fourier transform called a Fourier series.

11.1.3 Properties of transforms

The basis functions are usually chosen such that their spatial derivatives are proportional to the basis function itself,

$$\frac{d\varphi_m}{dx} = \alpha_m \varphi_m. \qquad (11.7)$$

The time and space derivatives of the transforms then have the following properties:

$$\frac{\partial}{\partial t} f(x,t) = \sum_{m=0}^{\infty} \frac{dF_m}{dt} \varphi_m \tag{11.8}$$

$$\frac{\partial}{\partial x} f(x,t) = \sum_{m=0}^{\infty} \alpha_m F_m \varphi_m. \tag{11.9}$$

These properties are what make transforms so useful to spectral modeling, since through their use we can convert a partial differential equation in space and time into an ordinary differential equation in time only.

11.2 The Spectral Method

Any basis functions that are orthogonal on the interval (a, b) may be used as basis functions for the Galerkin method. If the basis functions are continuous functions, then the method is also called a *spectral method*. If the basis functions are discontinuous or piecewise, then it is called a *finite-element* method. Spectral methods are more commonly used in atmospheric modeling applications than are finite-element methods.

11.2.1 Description of the spectral method

In this section we illustrate the general spectral method using the 1D advection equation

$$\frac{\partial u}{\partial t} + c \frac{\partial u}{\partial x} = 0. \tag{11.10}$$

We first represent $u(x,t)$ as a sum of the orthogonal basis functions in the form of (11.2),

$$u(x,t) = \sum_{l=0}^{N-1} U_l(t) \varphi_l(x), \tag{11.11}$$

with the time-dependent spectral coefficients represented by $U_l(t)$. Notice that we are not using an infinite sum, but only a finite number, N, of spectral components in the representation of u. The series

(11.11) is therefore *truncated* and will not be an exact representation for u.

Substituting (11.11) into (11.10) and using the properties for the derivatives, (11.8) and (11.9), we get

$$\sum_{l=0}^{N-1} \left(\frac{dU_l}{dt} \varphi_l + c\alpha_l U_l \phi_l \right) = 0. \tag{11.12}$$

If we multiply (11.12) by the conjugate of a basis function φ_m^*, and then integrate over the interval (a, b) we get

$$\int_a^b \sum_{l=0}^{N-1} \frac{dU_l}{dt} \varphi_l \varphi_m^* dx + \int_a^b \sum_{l=0}^{N-1} c\alpha_l U_l \varphi_l \varphi_m^* dx = 0. \tag{11.13}$$

Since the basis functions are orthogonal the only nonzero terms within the integrals will be those where $l = m$. We are then left with a system of N independent ordinary differential equations for the spectral coefficients,

$$\frac{dU_m}{dt} + c\alpha_m U_m = 0 \quad m = 0, 1, 2, \cdots, N - 1. \tag{11.14}$$

We can solve (11.14) using finite differencing in time to find the values of the spectral coefficients U_m at the new time, and then reconstruct $u(x, t)$ at the new time using (11.11).

The spectral technique can also be applied to systems of equations. In Exercise 11.1 we show that the spectral technique applied to the 1D barotropic primitive equations with no mean flow or Coriolis,

$$\frac{\partial u}{\partial t} = -g\frac{\partial \eta}{\partial x} \tag{11.15}$$

$$\frac{\partial \eta}{\partial t} = -H\frac{\partial u}{\partial x}, \tag{11.16}$$

results in the following equations for the spectral coefficients,

$$\frac{dU_m}{dt} = -\alpha_m g P_m \tag{11.17}$$

$$\frac{dP_m}{dt} = -\alpha_m H U_m, \tag{11.18}$$

where P_m are the spectral coefficients for η.

11.2.2 Benefit of the spectral method

The spectral technique allows us to convert partial differential equations in *physical space* (meaning that they are written in terms of the spatial variables x, y, and z) into ordinary differential equations in *spectral space* (written in terms of the wave numbers k, l, and m). These ordinary differential equations are then integrated forward in time. By transforming the equations into spectral space there is no need to approximate the spatial derivatives with finite differencing. This eliminates any errors associated with the spatial finite differencing, and is the single biggest advantage of the spectral method.

11.3 Fourier Transforms

11.3.1 Complex exponentials as basis functions

The most commonly used basis functions for one and two-dimensional applications are complex exponentials. For one-dimensional applications these take the form of

$$\varphi_m(x) = e^{\iota k_m x} \tag{11.19}$$

where

$$k_m = 2\pi m / (b - a) \quad m = 0, 1, 2, 3, \cdots, \infty. \tag{11.20}$$

Using these basis functions in (11.2) results in the familiar Fourier series expansion of $f(x)$, and the spectral coefficients are then called the *Fourier coefficients*. Subsituting (11.19) into (11.7) establishes that

$$\alpha_m = \iota k_m. \tag{11.21}$$

Using this result, the spectral form of the 1D advection equation, (11.14), becomes

$$\frac{dU_m}{dt} + \iota c k_m U_m = 0, \tag{11.22}$$

while the spectral forms of the 1D barotropic primitive equations, (11.17) and (11.18), are

$$\frac{dU_m}{dt} = -\iota g k_m N_m \tag{11.23}$$

$$\frac{dN_m}{dt} = -\iota H k_m U_m, \qquad (11.24)$$

for $m = 0, 1, 2, \cdots, N - 1$.

11.3.2 The fast Fourier transform

Most programming languages have readily available functions for calculating the Fourier coefficients of a function. These use an algorithm called the *fast Fourier transform*, or *FFT*. For the purposes of modeling, the one-dimensional FFT can be thought of as a 'black box' that inputs a function u_I digitized on N grid points ($I = 0, 1, 2, 3, \cdots, N - 1$), and returns the N Fourier coefficients U_m ($m = 0, 1, 2, 3, \cdots, N - 1$). The wave numbers corresponding to the spectral coefficients are given by

$$k_m = 2\pi m / N d, \qquad (11.25)$$

where d is the grid spacing. The *inverse fast Fourier transform* (*IFFT*) is a function that takes the N Fourier coefficients U_m as input and returns the function u digitized on the N grid points. These relationships are illustrated as

$$U = \text{FFT}(u)$$
$$u = \text{IFFT}(U). \qquad (11.26)$$

We do not need to know the details of how the FFT works in order to make use of it for spectrally solving an equation. Appendix C gives a more detailed explanation for the interested reader.

11.3.3 Advection equation example

We now illustrate the steps for spectrally solving the 1D advection equation on a grid consisting of N data points, using the fast Fourier transform.

1. The first step is to take the initial value of u at time step $n = 0$, and use the FFT to convert it into the initial Fourier coefficients,

$$U^0_{0:N-1} = \text{FFT}\left(u^0_{0:N-1}\right). \qquad (11.27)$$

2. Next we calculate the wave numbers k_m corresponding to the Fourier coefficients using (11.25) or a built-in function that many software libraries have for returning these wave numbers.

3. Use a finite-difference form of (11.22) to integrate forward in time to find the future values of the Fourier coefficients. If the 3rd-order Adams-Bashforth scheme is used, the finite-difference form of (11.22) will be

$$U_m^{n+1} = U_m^n - (\iota c k_m \Delta t / 12)(23 U_m^n - 16 U_m^{n-1} + 5 U_m^{n-2}).$$
(11.28)

4. After the solution is advanced to the desired time level we then use the inverse FFT to recover the future values of the function u at the grid points,

$$u_{0:N-1}^n = \text{IFFT}\left(U_{0:N-1}^n\right).$$
(11.29)

It is very important to realize when programming these equations that u and U are both complex-valued, meaning they have both real and imaginary components, though the imaginary component of u will be zero.

Figure 11.1 compares the spectral solution of the 1D advection equation with the standard finite-difference solution using centered space differencing. Both solutions use the 3rd-order Adams-Bashforth scheme for the time differencing. The spectral solution does a much better job of preserving the shape of the original signal. The 'noise' that is seen in the finite-difference solution is all due to the approximation of the spatial derivative using finite-differencing. This noise is absent in the spectral version because there is no spatial finite-differencing needed.[1]

11.4 The Transform Method

So far we have illustrated the spectral technique using only linear equations. The inclusion of nonlinear terms presents more difficulty, which we illustrate using the continuity equation from the 1D nonlinear barotropic primitive equations,

$$\frac{\partial \eta}{\partial t} + u \frac{\partial \eta}{\partial x} = -H \frac{\partial u}{\partial x}.$$
(11.30)

---

[1]There will still be computational modes if multi-level time-differencing schemes are used in spectral models.

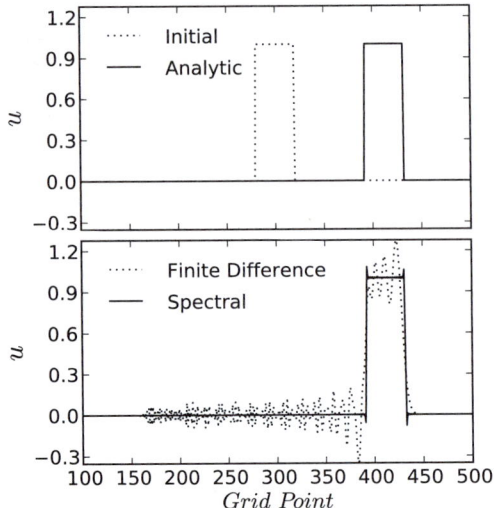

Fig. 11.1: The top panel shows a plot of u at the initial time and the analytic solution at a later time. The bottom panel shows the numerical solutions using centered finite differencing and also using the spectral technique. Both numerical solutions used 3rd-order Adams-Bashforth time differencing. The spectral solution is clearly superior to the finite-differenced solution.

To convert (11.30) into spectral form we write both u and η in terms of the basis functions,

$$u(x,t) = \sum_{l=0}^{N-1} U_l(t)\varphi_l(x) \tag{11.31}$$

$$\eta(x,t) = \sum_{n=0}^{N-1} P_n(t)\varphi_n(x), \tag{11.32}$$

and substituting into (11.30) obtain

$$\sum_{l=0}^{N-1} \frac{dP_l}{dt}\varphi_l + \left(\sum_{l=0}^{N-1} U_l\varphi_l\right)\left(\sum_{n=0}^{N-1} \alpha_n P_n \varphi_n\right)$$
$$= -H\left(\sum_{l=0}^{N-1} \alpha_l U_l \varphi_l\right). \tag{11.33}$$

If we multiply (11.33) by φ_m^* and then integrate the results of the multiplication, the result is

$$\int_a^b \sum_{l=0}^{N-1} \frac{dP_l}{dt} \varphi_l \varphi_m^* dx + \int_a^b \left(\sum_{l=0}^{N-1} U_l \varphi_l \right) \left(\sum_{n=0}^{N-1} \alpha_n P_n \varphi_n \right) \varphi_m^* dx$$

$$= - \int_a^b H \sum_{l=0}^{N-1} \alpha_l U_l \varphi_l \varphi_m^* dx, \quad (11.34)$$

which due to orthogonality becomes

$$\frac{dP_m}{dt} = -B_{mm}^{-1} \int_a^b \left(\sum_{l=0}^{N-1} U_l \varphi_l \right) \left(\sum_{n=0}^{N-1} \alpha_n P_n \varphi_n \right) \varphi_m^* dx$$

$$- \alpha_m H U_m. \quad (11.35)$$

Equation (11.35) looks just like (11.18) only with the additional, complicated term involving the product of two summations. This is problematic because each summation must be expanded into its constituent $N-1$ terms, and then the two summations must be multiplied together, which requires $(N-1)^2$ multiplication steps. Such an operation is very computationally intensive. In practice, what is done instead is to recognize that the two summations in (11.35) are themselves the inverse transforms of U_l and $\alpha_n P_n$ respectively,

$$u = \sum_{l=0}^{N-1} U_l \varphi_l \quad (11.36)$$

$$r = \sum_{n=0}^{N-1} (\alpha_n P_n) \varphi_n, \quad (11.37)$$

and that the middle term in (11.35) is just the mth spectral coefficient of the product of u and r,

$$Q_m = B_{mm}^{-1} \int_a^b u r \varphi_m^* dx. \quad (11.38)$$

Equation (11.35) then takes the form

$$\frac{dP_m}{dt} = -Q_m - H \alpha_m U_m. \quad (11.39)$$

To evaluate the middle term in (11.35) the steps are:

1. Take the forward transforms of u and η to get their spectral coefficients U and P.

2. Multiply P by α and take the inverse transform of the product to get the function r.

3. Multiply u and r together at each grid point to get the product ur on the grid.

4. Take the forward transform of ur to get their spectral coefficients Q.

5. Use the finite-difference form in time of (11.39) to find the future value of P.

6. Take the inverse transform of P to get the new value of η.

Even though the steps above are involved, and there is one additional forward and backward transform required, it still involves fewer multiplications than multiplying the two original series together term by term.

The method outlined above is referred to as the *transform method*. It is not strictly either a purely spectral method, or a purely grid-point method, but is a hybrid between the two, since at each time step both the spectral coefficient values and the values of the functions at the grid points are used. However, the transform method is used in most spectrally-based models, and such models are still commonly referred to as spectral models.

11.5   The Pseudospectral Method

The primary benefit of the spectral method is that the spatial derivatives do not need to be approximated using finite differencing. It is possible to use another hybrid method that solves the finite-difference equations at the grid points, but calculates any spatial derivatives in spectral space. This technique is known as the *pseudospectral* method.

The pseudospectral technique makes use of the following relationship for the spatial derivative of a generic variable, f,

$$\frac{\partial f}{\partial x} = \sum_{k=0}^{N-1} F_k \frac{d\varphi_k}{dx} = \sum_{k=0}^{N-1} \alpha_k F_k \varphi_k = \sum_{k=0}^{N-1} G_k \varphi_k. \tag{11.40}$$

We can therefore calculate the spatial derivative of f by the following steps:

1. Take the forward transform of f to find the spectral coefficients F_k.

2. Multiply the spectral coefficients F_k by α_k and call this new result G_k.

3. Take the inverse transform of G to recover the spatial derivatives at the grid points.

Using this method we can obtain the value of the spatial derivatives at the grid points without using finite differences.

11.6   Spherical Domains

Up until now we have assumed that our model domain is 1D. In real-world applications the spectral method must be applied in two horizontal dimensions on a spherical Earth.[2] The spherical grid dictates that different basis functions, called spherical harmonics, be used for spectral applications. The spherical grid also introduces two additional concerns: 1) the wind components are undefined at the poles; and 2) the longitude lines converge at the poles, so that the grid points become closer together as the poles are approached, requiring that very small time intervals be used to meet the CFL stability condition. The spectral method overcomes both of these difficulties.

The complete details of a fully spectral, global primitive equation model are beyond the scope of this introductory text. However, the following sections describe a few necessary details for implementing a barotropic primitive equation spectral model on a sphere, and provide a basis by which an interested reader may access more advanced treatments.

---

[2]In 3D models the spectral method is usually used only for the horizontal dimensions. The vertical dimension is still handled using finite differencing.

11.6.1   Spherical harmonics

The basis functions for spectral modeling on a spherical domain are called *spherical harmonics*, and are defined as

$$Y_{m,n}(\mu, \lambda) = \beta_{m,n} P_{m,n}(\mu) e^{\iota m \lambda}, \qquad (11.41)$$

where μ is the sine of the latitude ($\mu = \sin\phi$), λ is longitude, $P_{m,n}$ is the *associated Legendre function of the first kind* having *order m* and *degree n*, and $\beta_{m,n}$ is a *normalization factor*. The choice of normalization factor is arbitrary and varies among scientific and mathematical disciplines. The atmospheric sciences community generally uses a normalization factor of

$$\beta_{m,n} = \sqrt{\frac{(2n+1)(n-m)!}{(n+m)!}}, \qquad (11.42)$$

and so spherical harmonics are more commonly written as

$$Y_{m,n}(\mu, \lambda) = \sqrt{\frac{(2n+1)(n-m)!}{(n+m)!}} P_{m,n}(\mu) e^{\iota m \lambda}. \qquad (11.43)$$

The associated Legendre functions of the first kind are given by the formula[3]

$$P_{m,n}(\mu) = \frac{(1-\mu^2)^{m/2}}{2^n n!} \cdot \frac{d^{m+n}}{d\mu^{m+n}} (\mu^2 - 1)^n. \qquad (11.44)$$

---

[3]Caution! There are many alternate definitions and notations for Legendre functions and spherical harmonics in the literature. Different normalization factors are also used by different disciplines and software packages. Some sources also include the normalization factor $\beta_{m,n}$ in the definition of the associated Legendre function itself. It is also common to see the order written as a superscript and the degree as a subscript, and the function defined as $P_n^m(\mu) = (-1)^m \frac{(1-\mu^2)^{m/2}}{2^n n!} \cdot \frac{d^{m+n}}{d\mu^{m+n}}(\mu^2 - 1)^n$. The factor $(-1)^m$ is called the *Condon-Shortley phase*. In the atmospheric sciences we typically use the definition $P_{m,n}(\mu) = (-1)^m P_n^m(\mu)$, which yields (11.44). It is always important to understand the conventions and definitions of whichever reference is being consulted. For this text we are using notation and definitions consistent with Abramowitz, M. and Stegun, I. A., 1965: Legendre functions. *Handbook of Mathematical Functions*, Abramowitz, M. and I. A. Stegun, Eds., Dover, 1046 pp.

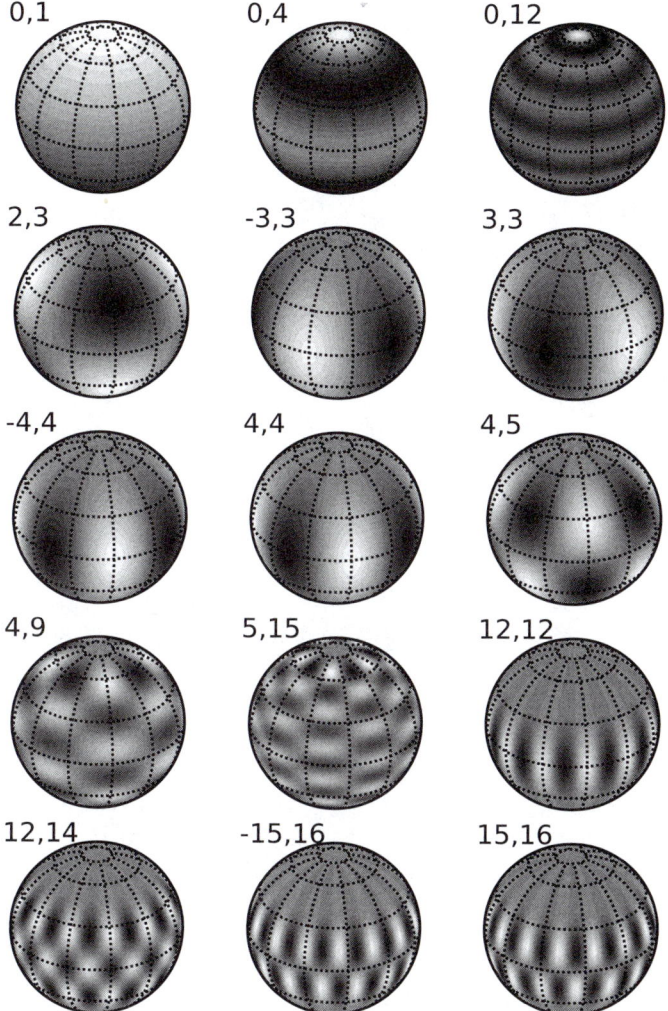

Fig. 11.2: Structure of the real part of spherical harmonics plotted on the globe. The number pairs are in the format *m, n*, where *m* is the order and *n* is the degree. Light shades represent positive values, while dark shades are negative. Notice that for even order the patterns for +*m* and −*m* are the same, while for odd order the patterns for +*m* and −*m* are opposite. The observation point is directly over latitude 30N and longitude 100W.

Figure 11.2 illustrates the structure of spherical harmonics of various order and degree. The order m may be either positive or negative, and its absolute value corresponds to how many nulls there are in one complete circle along a line of latitude. The difference between the degree and order, $n - |m|$, indicates how many nulls there are in one complete circle along a meridian. The degree must always be equal to or greater than the absolute value of the order, $n \geq |m|$.

Just as a function in a 1D or 2D Cartesian domain may be decomposed into a sum of complex exponential basis functions, a function $h(\mu, \lambda)$ on the surface of a sphere can be decomposed into a sum of spherical harmonic basis functions via the infinite series

$$h(\mu, \lambda) = \sum_{m=-\infty}^{\infty} \sum_{n=|m|}^{|m|+\infty} H_{m,n} Y_{m,n}, \quad (11.45)$$

where $H_{m,n}$ are the spectral coefficients. For modeling applications the series is truncated to a finite number of terms,

$$h(\mu, \lambda) \cong \sum_{m=-M}^{M} \sum_{n=|m|}^{|m|+M} H_{m,n} Y_{m,n}. \quad (11.46)$$

Some useful properties of spherical harmonics are that a harmonic of order and degree m, n can be rewritten in terms of harmonics of other orders and degrees via the recursion relations

$$Y_{m+1,n}(\mu, \lambda) = \frac{e^{\iota\lambda}}{(1-\mu^2)^{1/2}} \left[(n-m)\mu \left(\frac{\beta_{m+1,n}}{\beta_{m,n}} \right) Y_{m,n}(\mu, \lambda) \right.$$
$$\left. - (n+m) \left(\frac{\beta_{m+1,n}}{\beta_{m,n-1}} \right) Y_{m,n-1}(\mu, \lambda) \right] \quad (11.47)$$

$$(n-m+1)Y_{m,n+1}(\mu, \lambda) = (2n+1)\mu \left(\frac{\beta_{m,n+1}}{\beta_{m,n}} \right) Y_{m,n}(\mu, \lambda)$$
$$- (n+m) \left(\frac{\beta_{m,n+1}}{\beta_{m,n-1}} \right) Y_{m,n-1}(\mu, \lambda) \quad (11.48)$$

$$Y_{m,n+1}(\mu,\lambda) = \left(\frac{\beta_{m,n+1}}{\beta_{m,n-1}}\right) Y_{m,n-1}(\mu,\lambda)$$

$$- e^{\iota\lambda}(2n+1)(1-\mu^2)^{1/2}\left(\frac{\beta_{m,n+1}}{\beta_{m-1,n}}\right) Y_{m-1,n}(\mu,\lambda). \quad (11.49)$$

These three relations can be combined in various ways to derive other possible recursion relations between harmonics of various orders and degrees. The spatial derivatives of a spherical harmonic of order and degree m and n can also be written in terms of spherical harmonics of other orders and degrees, via

$$\frac{\partial Y_{m,n}}{\partial\lambda} = \iota m Y_{m,n} \quad (11.50)$$

and

$$(\mu^2 - 1)\frac{\partial Y_{m,n}}{\partial\mu} = \frac{n(n-m+1)}{2n+1}\frac{\beta_{m,n}}{\beta_{m,n+1}}Y_{m,n+1}$$

$$- \frac{(n+m)(n+1)}{2n+1}\frac{\beta_{m,n}}{\beta_{m,n-1}}Y_{m,n-1}. \quad (11.51)$$

This is reminiscent of the fact that the spatial derivatives of Fourier basis functions are written in terms of Fourier basis functions. Although these expressions are somewhat involved, the end result is that by writing all the dependent variables in terms of spherical harmonics and substituting these into the governing equations in spherical coordinates, the system of partial differential equations is converted into a system of ordinary differential equations in time for the spectral coefficients.

11.6.2 Streamfunction and velocity potential

The vector forms of the barotropic primitive equations are

$$\frac{\partial \vec{V}}{\partial t} + \vec{V}\cdot\nabla\vec{V} = -g\nabla\eta - \hat{k}\times f\vec{V} \quad (11.52)$$

$$\frac{\partial\eta}{\partial t} + \nabla\cdot\left(\eta\vec{V}\right) = -H\nabla\cdot\vec{V}. \quad (11.53)$$

The vector forms of the equations are not tied to any particular coordinate system. It is only when we write the equations in component form that we must account for the type of coordinate system used (e.g., Cartesian, polar, spherical, etc.) The dependent variables in (11.52) and (11.53) are u, v, and η.

The first step toward using streamfunction and velocity potential as dependent variables is to derive the vorticity and divergence equations from (11.52). The vorticity equation is derived by applying the vector operator $\hat{k} \cdot \nabla \times$ to (11.52), the result of which is

$$\frac{\partial \zeta}{\partial t} + \nabla \cdot (\zeta + f) \vec{V} = 0. \tag{11.54}$$

The divergence equation is derived by applying the operator $\nabla \cdot$ to (11.52) which results in

$$\frac{\partial \delta}{\partial t} - \hat{k} \cdot \left[\nabla \times (\zeta + f) \vec{V} \right] + \nabla^2 \left(\frac{\vec{V} \cdot \vec{V}}{2} \right) + g\nabla^2 \eta = 0. \tag{11.55}$$

The continuity equation can be written as

$$\frac{\partial \eta}{\partial t} + \nabla \cdot \left(\eta \vec{V} \right) = -H\delta. \tag{11.56}$$

Recalling that vorticity can be written in terms of streamfunction as

$$\zeta = \nabla^2 \psi, \tag{11.57}$$

divergence can be written in terms of velocity potential as

$$\delta = \nabla^2 \chi, \tag{11.58}$$

and also that $\vec{V}$ can be written in terms of both streamfunction and velocity potential as

$$\vec{V} = \hat{k} \times \nabla \psi + \nabla \chi, \tag{11.59}$$

equations (11.54), (11.55), and (11.56) can be written entirely in terms of the dependent variables ψ, χ, and η.

Exercises

Ex. 11.1:   Show that the general spectral technique applied to the 1D barotropic primitive equations, (11.15) and (11.16), results in (11.17) and (11.18).

Ex. 11.2: Carry out the steps leading from (11.30) to (11.35).

Ex. 11.3: Write a program to solve the 1D advection equation using the spectral method with 3rd-order Adams-Bashforth time differencing. Use complex-exponential basis functions on a grid containing 601 grid points and a wave speed of 15 m s^{-1}, a grid spacing of 1 km, and a time increment of 5 s. Initialize the solution with a box of width 40 grid points, centered on the grid. The results after 1500 iterations should look similar to the numerical solution in the bottom panel of Fig. 11.2.

## CHAPTER 12

Data Assimilation

12.1 Introduction

12.1.1 Importance

Atmospheric modeling is at its core an initial value problem. The solution (the state of the atmosphere) at some time $t = \tau$ is dependent on the initial state at $t = 0$, plus the integrated effects of all the forcing terms applied during the time period from zero to τ. If the initial conditions are not representative of the actual state of the atmosphere at $t = 0$ there is little hope of obtaining an accurate solution at a later time.

The initial state of the atmosphere is represented by observations taken from a variety of platforms and methods. Surface observations taken from regular reporting stations are certainly a primary data source, but only represent conditions at ground level. Upper air observations from radiosondes are taken by far fewer observing stations and only twice daily. Satellite observations of temperature, humidity, and wind are able to fill in many data gaps, as do observations from commercial and military aircraft, radars, acoustic profilers, lidar, and other remote sensing methods.

Even if there are abundant observations available for use as initial conditions they are going to be irregularly spaced, and will not have all been made at the exact same time. The observations must therefore be interpolated to the model grid and also adjusted temporally to a common time. Even so, there will still be many regions of the atmosphere, particularly over remote oceans with little ship or aircraft traffic, where there will be data voids. Short-term forecasts from previous model runs are sources of information that are readily available and used to fill these voids.

Interpolating the observations to the model grid points at a common initialization time and blending them with the short-term model forecast is certainly one important aspect of model initialization, but it is only half the problem. Recall from Chapter 6 that L. F. Richardson's early attempt at numerical prediction was a failure in large part because the initial conditions that he used were not balanced. If the actual atmosphere is pushed out of balance, the response is the generation of gravity waves which redistribute mass and momentum. Even small departures from balance will result in the generation of gravity waves. If the model initial conditions are out of balance, the simulation will result in large-amplitude gravity waves that dominate the solution and swamp the small but meteorologically meaningful pressure changes.

Model initialization therefore has two very important functions: 1) Interpolate all available observations to the model grid at a common initialization time, and 2) Balance the initial conditions to avoid the generation of spurious, large-amplitude gravity waves in the solution.

12.1.2   A brief history

The history of data assimilation dates back to the late 1940s, when the first simple data analysis based on interpolation methods and called *objective analysis* was developed. The methods were called *objective* because they were based on mathematical relationships, as compared to *subjective analyses* that involved manipulation of the data by human intervention. All data assimilation methods for numerical models are now objective techniques, but the term objective analysis is now often reserved for denoting a specific class of simple interpolation techniques.

During the 1950s and early 1960s, objective initialization techniques evolved for operational purposes. Examples of methods that were developed by meteorologists, such as George Cressman and Stanley Barnes, are based upon iterative or *successive correction* methods. The idea of using a 'first guess', or background field was also introduced, using a short-term forecast from a prior model run in order to fill in data poor regions. A major advancement in the field of data assimilation occurred in the early 1960s, when Lev Gandin introduced a statistical least squares interpolation method based upon earlier ideas of Andrei Kolmogorov.[1] This method is now known as *optimal interpolation* or simply *OI*.

All of the early data assimilation methods were static in their approach with respect to time, using only data valid at a single time. In the 1970s, a data assimilation approach called *Newtonian relaxation* or *nudging* was developed that allowed the observed data to be assimilated by inserting artificial forcing terms in the governing equations over a finite period of time at the beginning of the simulation, to gently nudge the model solution toward convergence with the observed data.

Since the 1980s, as computational capabilities have increased, more complex and computationally intensive data analysis and assimilation schemes have been developed. These include three- and four-dimensional variational methods (3DVAR and 4DVAR) and the ensemble Kalman filter (EnKF). The field of data assimilation is still rapidly evolving. The remainder of this chapter will concentrate on the methodology and schemes used in the data analysis, assimilation, and initialization steps.

12.2 Data Assimilation, Data Analysis, and Initialization

Data assimilation describes the entire process whereby the initial fields for a model simulation are constructed. The data assimilation process is comprised of distinct steps, shown in Fig. 12.1. The two most critical of these steps are *data analysis* and *initialization*.

12.2.1 Data analysis

The objective of the data analysis step is to produce the most accurate possible representation of the actual state of the atmosphere at

---

[1]Gandin, L.S., 1963: *Objective Analysis of Meteorological Fields*, Israel Program for Scientific Translations, 242 pp.

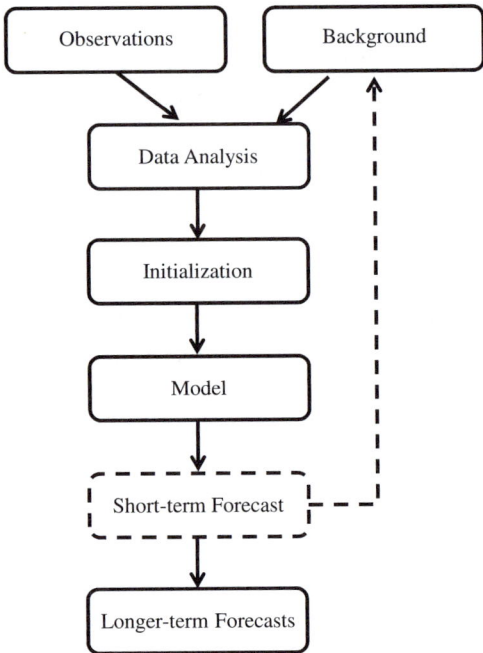

Fig. 12.1: The data assimilation cycle. The background field can come from climatology or output from a larger-scale model (warm start), or as represented by the dashed lines a short-term forecast from a previous run of the same model (cold start).

a given time on the model grid. The results from the data analysis step can be useful as diagnostic tools for investigating atmospheric structure and processes and can also be used as input to other models or analysis programs.

In an ideal world there would be simultaneous, error-free observations of all relevant atmospheric variables taken at the locations of every model grid point, and these would be used as the initial conditions for the model simulation. In the real world the observations are sparse and are not collocated with the model grid points. The observations are also not necessarily taken simultaneously. There are errors associated with instrumentation, human reading of instruments, and data precision.

The data analysis step takes all of these imperfect, irregularly spaced observations and attempts to create a regularly-spaced gridded data set for use in the model simulation. Since there are large areas of the globe with a dearth of observations, additional information must also be incorporated into most simulations. This additional information is often in the form of a *background* or *first-guess* field. The background field may be produced from climatology or output from a larger-scale model, in which case the assimilation process is referred to as a *cold-start*. The background field may also be produced using a short-term forecast from a previous run from the same model, in which case the assimilation process is referred to as a *warm-start*.

Another additional source of data are *synthetic* or *bogus observations* that are artificially generated to represent known circulations such as tropical cyclones, for which there are not adequate, real observations available.

In the data analysis phase the various data sources (observations, background field, synthetic observations) will have different weighting to determine how much each will influence the final, analyzed fields at a particular grid point. The differences in the various data analysis techniques often lie in what these weightings are and how they are applied. The data analysis phase will also account for whether the values at the model grid points represent model conditions at the specific grid point, or the average values for an entire grid volume.

12.2.2   Initialization

The gridded output from the data analysis step cannot be used directly as initial conditions for a primitive equation model. This is because it is unlikely that the kinematic variables (u, v, and w) and the thermodynamic variables (T, p, ρ, and q) will be in proper balance, and so once the simulation begins there will be spurious gravity waves generated as the simulation attempts to restore balance. The objective of the initialization step is to create conditions that are in balance. Paradoxically, because of the requirement for balanced initial conditions, the best initial conditions for the model simulation may not necessarily be the most accurate conditions representing the actual state of the atmosphere.

12.2.3   Categories and techniques

In the remainder of this chapter we examine some selected methods and procedures for data assimilation that are representative of the various types of methods available. The methods we discuss are: 1) successive correction methods, illustrated by the venerable Cressman and Barnes schemes; 2) least squares methods, illustrated by the optimal interpolation and Kalman filtering methods; and 3) variational methods. Our treatment is not meant to be exhaustive, but is instead intended to give a very cursory introduction serving as a bridge to more comprehensive and advanced treatments.[2]

12.3   Successive Correction Methods

In successive correction methods the field variables are modified by the observations in an iterative manner. A pass is made through every grid point, updating the variable at the point based on the background field and the observations surrounding the grid point. After one pass is made through the domain, another pass is made, again modifying the field at each grid point based on the observations surrounding the grid point.

If there are a total of K observation points labeled as $O_1, O_2, O_3, \cdots, O_K$, then the value of the variable s at grid point i is updated according to the formula

$$s_i^{m+1} = s_i^m + \frac{\sum\limits_{k=1}^{K} w_{ik}^m \left(O_k - s_k^m \right)}{\sum\limits_{k=1}^{K} w_{ik}^m + \varepsilon^2} \tag{12.1}$$

where s_i^m is the value of the field (e.g., T, q, u, etc.) at the ith grid point at the mth iteration, s_k^m is the value of the field at observation point k (interpolated from the gridded field) at the mth iteration, w_{ik}^m is a weighting function which depends on how far the observation is from the grid point, and ε^2 is an estimate of the ratio of the observation error to the first guess field error (if the observations were perfect, then $\varepsilon^2 = 0$).

---

[2]Chapter 5 of Kalnay, E., 2003: *Atmospheric Modeling, Data Assimilation and Predictability*, Cambridge, 341 pp. provides a reasonably accessible exposition of modern data assimilation techniques.

The two main successive correction schemes that are used are the Cressman and the Barnes schemes. Both schemes use (12.1), and differ only in how the weighting function w is defined.

12.3.1 Cressman scheme

In the Cressman scheme the weighting function is defined in terms of a *radius of influence*, R_m, using the formula

$$
w_{ik}^m = \begin{cases} \frac{R_m^2 - r_{ik}^2}{R_m^2 + r_{ik}^2}, & \text{for } r_{ik}^2 \leq R_m^2; \\ 0, & \text{for } r_{ik}^2 > R_m^2; \end{cases} \tag{12.2}
$$

where r_{ik}^2 is the distance from the observation to the grid point. In this scheme, observations outside the radius of influence have no bearing on the result. Those observations within the radius of influence are weighted according to how close they are to the grid point, with the closer observations receiving a greater weight.

Cressman's scheme assumes that the observations are perfectly error-free ($\varepsilon^2 = 0$). The radius of influence is usually set to decrease for each pass through the field, so that the field is corrected to larger-scale features during the first iterations, and to smaller-scale features during later iterations. Four to six iterations are commonly used.

12.3.2 Barnes scheme

The Barnes scheme is a widely used method that has largely replaced the Cressman scheme. The Barnes scheme also uses (12.1) and assumes that the observations are error-free, but defines the weighting function as

$$
w_{ik}^m = \exp\left(-\frac{r_{ik}^2}{4c_m}\right). \tag{12.3}
$$

The parameter c_m controls the rate of fall-off of the weighting function with distance from the grid point. Since (12.3) is asymptotic to zero as distance increases, there is no need to specify a radius of influence. For each successive iteration the parameter c_m becomes smaller according to

$$
c_{m+1} = \alpha c_m \tag{12.4}
$$

where $0 < \alpha < 1$. This results in the weighting having a steeper fall-off with distance for each successive iteration, so like in the Cressman scheme, the first pass through the data is weighted toward larger-scale features, while subsequent passes capture smaller-scale features. Most applications of the Barnes scheme use only two passes through the data.

The advantages of the Barnes scheme are: 1) the degree of smoothing in the analysis can be predetermined and selectively set, depending on the situation; 2) small-scale noise is suppressed so that further smoothing with numerical filters is not necessary; 3) since the weighting function approaches zero asymptotically, the influence of the observations may be extended to any distance without changing the weighting function and response characteristics; 4) desired scale resolution is achieved with only one iteration, so only two passes are necessary; 5) no background field is required; 6) time-weighting of observations is possible.

12.3.3 Disadvantages

The Cressman and Barnes schemes have some common disadvantages. Neither is suitable for diverse observations. This is because observational error is not accounted for when determining how observations are weighted. An observation from an *in situ* instrument would be given the same weight as a remotely sensed observation.

Another problem is that the weighting does not depend on the density of observations. For example, if there are five very closely-spaced observations, and a sixth observation that is well away from the others, the isolated observation provides more independent information than any of the individual, closely spaced observations. However, in either the Cressman or Barnes schemes, all six observations would be given the same weight, skewing the resulting analysis toward the cluster of observations.

Finally, neither the radius of influence R_m in the Cressman scheme, nor the scaling factor c in the Barnes scheme, are based on any physical or statistical properties of the background state or the observations.

12.4   Least Squares Methods

Least squares methods differ from successive correction methods in that the observations are weighted according to some known or estimated statistics regarding their errors, rather than empirical values. Thus, observations from different sources are weighted differently based on known instrument and other errors. For example, radiosonde measurements of temperature can be given a greater weight than satellite-derived temperatures.

12.4.1   The analysis equation

Least squares methods attempt to minimize the total error of all the observations to come up with an 'ideal' weighting for the observations. With seven different variables and thousands of model grids and observations, the process can become quite complex. However, the basic concept can be illustrated using a gross oversimplification, analyzing only three fields of temperature (T), pressure (p) and humidity (q) at two grid points, using only two observations. The analysis values of these fields at a grid point will be a linear combination of the observations and the background field, given by the equations

$$T_{a1} = T_{b1} + k_{T11}(T_{O1} - \tilde{T}_{b1}) + k_{T12}(T_{O2} - \tilde{T}_{b1})$$

$$p_{a1} = p_{b1} + k_{p11}(p_{O1} - \tilde{p}_{b1}) + k_{p12}(p_{O2} - \tilde{p}_{b1})$$

$$q_{a1} = q_{b1} + k_{q11}(q_{O1} - \tilde{q}_{b1}) + k_{q12}(q_{O2} - \tilde{q}_{b1})$$

$$T_{a2} = T_{b2} + k_{T21}(T_{O1} - \tilde{T}_{b2}) + k_{T22}(T_{O2} - \tilde{T}_{b2})$$

$$p_{a2} = p_{b2} + k_{p21}(p_{O1} - \tilde{p}_{b2}) + k_{p22}(p_{O2} - \tilde{p}_{b2})$$

$$q_{a2} = q_{b2} + k_{q21}(q_{O1} - \tilde{q}_{b2}) + k_{q22}(q_{O2} - \tilde{q}_{b2}).$$

$$(12.5)$$

The convention in (12.5) is that S_{an} is the analysis value of the field S (e.g., T, p, q, u, v, w) at grid point n, S_{bn} is the background value of the field at grid point n, S_{Om} is the observed value of the field at observation point m, and $\tilde{S}_{bm}$ is the background value of the field at the observation point m. It is important to note the distinction between S_{bn} and $\tilde{S}_{bm}$; they are both values from the background field, but the ones without the tilde are valid at the model grid points, while the ones with the tilde are valid at the observation points. The weighting coefficients k_{Snm} are interpreted as the weighting for field S at grid point n of the observation at point m.

Equation (12.5) can be written in matrix form as

$$\mathbf{x}_a = \mathbf{x}_b + \mathbf{K}\mathbf{d}, \tag{12.6}$$

with the vectors and matrices defined as the *analysis vector*,

$$\mathbf{x}_a = \begin{pmatrix} T_{a1} \\ p_{a1} \\ q_{a1} \\ T_{a2} \\ p_{a2} \\ q_{a2} \end{pmatrix}; \tag{12.7}$$

the *background vector*,

$$\mathbf{x}_b = \begin{pmatrix} T_{b1} \\ p_{b1} \\ q_{b1} \\ T_{b2} \\ p_{b2} \\ q_{b2} \end{pmatrix}; \tag{12.8}$$

the *gain* or *weight matrix*, which contains information about the errors associated with the observations and background fields,

$$\mathbf{K} = \begin{pmatrix} k_{T11} & k_{T12} & 0 & 0 & 0 & 0 & 0 & 0 & 0 & 0 & 0 & 0 \\ 0 & 0 & k_{p11} & k_{p12} & 0 & 0 & 0 & 0 & 0 & 0 & 0 & 0 \\ 0 & 0 & 0 & 0 & k_{q11} & k_{q12} & 0 & 0 & 0 & 0 & 0 & 0 \\ 0 & 0 & 0 & 0 & 0 & 0 & k_{T21} & k_{T22} & 0 & 0 & 0 & 0 \\ 0 & 0 & 0 & 0 & 0 & 0 & 0 & 0 & k_{p21} & k_{p22} & 0 & 0 \\ 0 & 0 & 0 & 0 & 0 & 0 & 0 & 0 & 0 & 0 & k_{q21} & k_{q22} \end{pmatrix}; \tag{12.9}$$

and the *observation increment*, which is just the difference between the observed values and the background values at the observation points,

$$
\mathbf{d} = \begin{pmatrix}
T_{O1} - \tilde{T}_{b1} \\
T_{O2} - \tilde{T}_{b1} \\
p_{O1} - \tilde{p}_{b1} \\
p_{O2} - \tilde{p}_{b1} \\
q_{O1} - \tilde{q}_{b1} \\
q_{O2} - \tilde{q}_{b1} \\
T_{O1} - \tilde{T}_{b2} \\
T_{O2} - \tilde{T}_{b2} \\
p_{O1} - \tilde{p}_{b2} \\
p_{O2} - \tilde{p}_{b2} \\
q_{O1} - \tilde{q}_{b2} \\
q_{O2} - \tilde{q}_{b2}
\end{pmatrix}. \tag{12.10}
$$

The observation increment can be rewritten as

$$
\mathbf{d} = \begin{pmatrix}
T_{O1} - \tilde{T}_{b1} \\
T_{O2} - \tilde{T}_{b1} \\
p_{O1} - \tilde{p}_{b1} \\
\vdots \\
q_{O1} - \tilde{q}_{b2} \\
q_{O2} - \tilde{q}_{b2}
\end{pmatrix} = \begin{pmatrix}
T_{O1} \\
T_{O2} \\
p_{O1} \\
\vdots \\
q_{O1} \\
q_{O2}
\end{pmatrix} - \begin{pmatrix}
\tilde{T}_{b1} \\
\tilde{T}_{b1} \\
\tilde{p}_{b1} \\
\vdots \\
\tilde{q}_{b2} \\
\tilde{q}_{b2}
\end{pmatrix} = \mathbf{y} - \mathbf{H}\mathbf{x}_b, \tag{12.11}
$$

where $\mathbf{y}$ is a vector containing all the observation values and $\mathbf{H}$ is the *linear observation operator matrix*,[3] which when multiplied with the background field vector $\mathbf{x}_b$ (at the model grid point values) returns a vector containing the background fields at the observation points. Thus, (12.6) is often written as

$$
\mathbf{x}_a = \mathbf{x}_b + \mathbf{K}\left(\mathbf{y} - \mathbf{H}\mathbf{x}_b\right). \tag{12.12}
$$

Equation (12.12) is called the *analysis equation*, and can be used for any number of grid points, observations, and fields.

12.4.2 Quantifying the errors

The errors in the analysis arise from: 1) errors in the background fields, or 2) errors in the observations. The *analysis error vector* is the

---

[3]In general the observation operator can be nonlinear, in which case it is not a matrix. The most general form of the observation increment is $\mathbf{d} = \mathbf{y} - H(\mathbf{x}_b)$, but in the linear case it is $\mathbf{d} = \mathbf{y} - \mathbf{H}\mathbf{x}_d$. The linear case is suitable as long as the departures of the observations from the background state are not large.

difference between the analysis values and the 'true' values of the fields (denoted by $\mathbf{x}_t$),

$$\varepsilon_a = \mathbf{x}_a - \mathbf{x}_t. \tag{12.13}$$

Multiplying the analysis error vector by its transpose, and then averaging, yields the *analysis error covariance matrix*,

$$\mathbf{A} = \overline{\varepsilon_a \varepsilon_a^\mathsf{T}}. \tag{12.14}$$

The *background error covariance matrix* is similarly constructed from the *background error vector*,

$$\varepsilon_b = \mathbf{x}_t - \mathbf{x}_b, \tag{12.15}$$

and is

$$\mathbf{B} = \overline{\varepsilon_b \varepsilon_b^\mathsf{T}}. \tag{12.16}$$

Likewise, the *observation error covariance matrix* comes from the *observation error vector*,

$$\varepsilon_o = \mathbf{y} - \mathbf{H}\mathbf{x}_t, \tag{12.17}$$

and is

$$\mathbf{R} = \overline{\varepsilon_o \varepsilon_o^\mathsf{T}}. \tag{12.18}$$

The elements of the error covariance matrices indicate the degree of correlation of the errors between grid points. If the errors between two grid points are completely uncorrelated, then the element of the error covariance matrix corresponding to these two grid points will be zero.

12.4.3   Optimum weighting coefficients

The analysis equation, (12.12), states that the analysis fields $\mathbf{x}_a$ are composed of the background fields, $\mathbf{x}_b$, plus corrections based on the differences between the observations $\mathbf{y}$ and the background fields at the observations points $\mathbf{H}\mathbf{x}_b$, appropriately weighted by the weighting coefficients contained in the gain matrix $\mathbf{K}$. If the observations and the background field were both error free, then the weighting coefficients would be functions solely of the distance between the observation point and the grid point. However, the observations and the background field will contain errors, so the optimal weighting coefficients will also be functions of these errors, or more appropriately, statistics about the errors.

The value for the optimal weighting matrix is determined from the matrix equation

$$\mathbf{K} = \mathbf{BH}^{\mathbf{T}}\left(\mathbf{HBH}^{\mathbf{T}} + \mathbf{R}\right)^{-1}. \qquad (12.19)$$

Equations (12.12) and (12.19) form the basis for the least squares methods of data analysis. In theory, if the observation and background error covariance matrices are known, then (12.19) is solved for the optimal weight matrix coefficients and then the latter are applied to (12.12) to find the analysis field.

The gist of least squares methods is to determine the optimum values for the elements of the weight or gain matrix, $\mathbf{K}$, such that the analysis error is minimized. There are two primary types of methods in use for this: 1) *optimal interpolation* and 2) *Kalman filtering*. Though both methods use (12.12), they differ as to how the background error covariance matrix $\mathbf{B}$ is estimated for use in (12.19).

In the method of optimal interpolation, the elements of the background error covariance matrix $\mathbf{B}$ are estimated based on long-term statistical comparisons between the background fields and analysis fields over many prior model runs. The error covariance matrix (or more appropriately, the method used to estimate the components of the matrix) is essentially static, using statistics built up over many prior model runs. This is very different from the Kalman filtering methods in which the error covariance matrix coefficients are dynamically recomputed and updated at every analysis time based on the departures between the previous analysis and previous background fields. So, while both methods are essentially tackling the same problem, and both use a short-term forecast as the background field, their methodology is very different in how the error covariance matrix is constructed. In most of the literature, the symbol $\mathbf{B}$ is reserved for the background error covariance matrix used in optimal interpolation, while the symbol $\mathbf{P}_f$ is used for that of the Kalman filtering method.[4]

In either the optimal interpolation or Kalman filtering methods it would be ideal to solve (12.12) and (12.19) simultaneously over the entire model domain. However, for a large-scale model with

---

[4]The subscript f denotes 'forecast' and serves to remind us that the Kalman filtering method calculates the error covariances based on the most recent short-term model forecast.

multiple model variables, grid points, and observations, the matrices involved are quite large, and beyond the practical ability to solve directly. Thus, the analysis is usually performed grid-point by grid-point or sequentially on limited subdomains of the grid.

12.5 Variational Methods

None of the methods discussed so far address the initialization step of the data assimilation process (refer to Fig. 12.1). The output fields from these objective analysis schemes are still likely to be unbalanced, and cannot be ingested directly into a primitive equation model. The variational techniques which we now discuss actually perform both the objective analysis and initialization steps at once.

Variational methods have their roots in the *calculus of variations*, which is a branch of mathematical analysis aimed at finding functions which minimize or maximize integrals or parameters. Variational methods are used in many branches of physics to solve problems such as finding the path of minimum energy for a particle to travel between two points in the presence of a potential field. Variational methods are powerful because any constraints imposed on the system, such as forcing a particle to remain on the surface of a cone or sphere, can be incorporated into the formulation of the problem before it is solved.

In the 3-dimensional variational approach (3DVAR) to data assimilation the property to be minimized is the cost function, defined as

$$J(\mathbf{x}) = \frac{1}{2} \left[(\mathbf{x} - \mathbf{x}_b)^\mathrm{T} \mathbf{B}^{-1} (\mathbf{x} - \mathbf{x}_b) + (\mathbf{y} - \mathbf{H}\mathbf{x})^\mathrm{T} \mathbf{R}^{-1} (\mathbf{y} - \mathbf{H}\mathbf{x}) \right]. \quad (12.20)$$

The value of $\mathbf{x}$ that results in the minimum value of the cost function $J(\mathbf{x})$ is the optimal analysis field. All terms and variables appearing on the right-hand side of (12.20) are as previously defined. It is unlikely that the problem can be solved analytically, and instead must be solved numerically using an iterative process. The procedure is often performed globally over the entirety of the model grid points, fields, and observations, instead of performing a series of piecemeal local analyses as is often done with optimal interpolation.

Though it looks very different than the optimal interpolation method, the mathematical formulation of the 3D variational method is actually formally equivalent to the optimum interpolation method. The fundamental difference lies in what entity is being solved for. In the optimal interpolation method we are solving for the optimal weight matrix, which is then used to determine the analysis field. In the variational approach we are solving for the analysis field directly.

The variational approach has several advantages over the optimal interpolation approach, though one main advantage is that it is possible to modify the cost function to include constraints such as gradient and hydrostatic balance. Then, a minimization of the cost function automatically returns fields that are balanced. Thus, the variational approach not only accomplishes the data analysis portion of the data assimilation process, but the initialization step as well.

The 3D variational method has also been extended to include observations that are not coincident in time. This approach, termed 4DVAR, is more computationally intensive than is 3DVAR.

12.6   Newtonian Relaxation

Another possible data assimilation scheme that is used for certain special applications is the Newtonian relaxation method. This is a dynamic initialization method that is more popularly called *nudging*. In this technique a nudging or forcing term is added to the prognostic equations for the field variables. For example, if the prognostic equation is of the form

$$\frac{\partial s}{\partial t} = F(s, \vec{V}, \cdots) \tag{12.21}$$

and there is an observed value of s, denoted as s_O, then a nudging term of the form $(s_O - s)/\tau_s$ is added to the prognostic equation. The model equation is then

$$\frac{\partial s}{\partial t} = F(s, \vec{V}, \cdots) + (s_O - s)/\tau_s. \tag{12.22}$$

When the equations are then integrated forward, the result of the nudging term is to force the solution to converge toward the observed value, s_O. The parameter τ_s is the *relaxation time*. Nudging does have an advantage of also balancing the initial fields, since as the model equations are integrated forward they will undergo adjustment.

12.7 Initialization

Most operational models, and many research models, now use variational techniques for data assimilation. Since variational techniques are able to accomplish the data analysis and initialization steps at the same time, the output fields from these techniques are already in balance and ready to input into a model. Prior to the widespread use of variational techniques, the initialization step had to be performed after the analysis step.

There are several methods that have been used or are still in use for performing the initialization step. The most venerable of these techniques is the *nonlinear normal-mode initialization*, which was the most used technique prior to variational methods. Another technique, that is still in use, is the implementation of digital filters to smooth out the perturbations caused by the generation of gravity waves. The details of these techniques are beyond the scope of this text.

## CHAPTER 13

Primitive Equation Models

13.1 General

Until now we have focused on modeling techniques and strategies that avail themselves to examples that can be programmed relatively easily. The full set of primitive equations contains seven equations and seven unknowns, and requires detailed and complex parameterizations for subgrid-scale phenomena. In concept it is fairly straightforward to program a full primitive equation model, but actually doing so would require a vast amount of code with many separate modules for handling parameterizations for turbulence, clouds, precipitation, radiation, latent heat release, and interactions between the atmosphere and the surface of the Earth. We would quickly get bogged down in details when trying to describe how such models are numerically integrated. Instead, this chapter presents a qualitative discussion of the differences in the general types of primitive equation models and how they are implemented.

The use of the primitive equations presents several choices, the first of which is whether to use the full vertical momentum equation (nonhydrostatic models) or the hydrostatic approximation. Another choice is which of the three forms of the continuity equation to use: compressible, anelastic, or incompressible. These choices are dictated both by the scale of the phenomenon to be modeled and by computational performance requirements.

13.2 The Primitive Equations

The full set of primitive equations in Cartesian height coordinates, including all Coriolis terms, is

$$\frac{Du}{Dt} = -\frac{1}{\rho}\frac{\partial p}{\partial x} + 2\Omega \sin \phi\, v - 2\Omega \cos \phi\, w + F_x \qquad (13.1)$$

$$\frac{Dv}{Dt} = -\frac{1}{\rho}\frac{\partial p}{\partial y} - 2\Omega \sin \phi\, u + F_y \qquad (13.2)$$

$$\frac{Dw}{Dt} = -\frac{1}{\rho}\frac{\partial \rho}{\partial z} - 2\Omega \cos \phi\, u - g + F_z \qquad (13.3)$$

$$\frac{\partial \rho}{\partial t} + \nabla \cdot \left(\rho \vec{V}\right) = 0 \qquad (13.4)$$

$$c_p \frac{DT}{Dt} - \frac{1}{\rho}\frac{Dp}{Dt} = \dot{Q} + F_T \qquad (13.5)$$

$$p = \rho R_d T \left(1 + 0.61q\right), \qquad (13.6)$$

and

$$\frac{\partial(\rho q)}{\partial t} + \nabla \cdot (\rho q \vec{V}) = S_O - S_K + F_q. \qquad (13.7)$$

By way of review, these equations are:

- The horizontal momentum equations, (13.1) and (13.2). The terms F_x and F_y represent the turbulent fluxes of horizontal momentum and must be parameterized.

- The vertical momentum equation, (13.3). The term F_z represents turbulent fluxes of vertical momentum.

- The continuity equation in fully-compressible form, (13.4).

- The thermodynamic energy equation, (13.5). The term $\dot{Q}$ represents diabatic heating via processes such as radiation and latent heating and cooling. The term F_T represents turbulent heat fluxes. Terms $\dot{Q}$ and F_T must be parameterized.

- The equation of state (ideal gas law), (13.6).

- The water-mass continuity equation, (13.7). The terms S_O and S_K represent sources and sinks of water vapor, while the term F_q represents turbulent moisture fluxes. Terms S_O, S_K and F_q must be parameterized.

13.3 Vertical Pressure Balance

The full set of primitive equations support numerous wave modes as solutions. Some of these waves are meteorologically significant and need to be retained for optimal model performance. Among these are the Rossby waves and inertio-gravity waves. However, the equations also support acoustic waves, which are not meteorologically significant. Since acoustic waves are fast-moving, they place a limitation on the time interval and grid spacing of the model. This limitation is most severe in the vertical direction, since the vertical grid spacing is usually much smaller than the horizontal grid spacing. If we can eliminate vertically propagating acoustic waves, then a longer time interval can be used in the model.

13.3.1 Hydrostatic models

Vertically propagating acoustic waves can be eliminated by using the hydrostatic approximation, in which the vertical momentum equation, (13.3), is substituted with the hydrostatic equation,

$$\frac{\partial p}{\partial z} = -\rho g. \tag{13.8}$$

Not only does this allow a longer time interval to be used, since vertically propagating acoustic waves are eliminated, but it also simplifies the equations considerably. It does, however, also mean that we no longer have a prognostic equation for vertical velocity, which instead must be diagnosed from another equation that contains w as a dependent variable.[1]

---

[1]In general there are three methods, each having variants, for diagnosing vertical velocity: (1) the *kinematic method*, which involves integration of the continuity equation; (2) the *thermal method*, which uses the thermodynamic energy equation; and (3) the *vorticity method*, which uses the quasigeostrophic vorticity equation.

13.3.2 Nonhydrostatic models

As the horizontal grid spacing decreases, the use of the hydrostatic approximation becomes less valid. Though some larger-scale mesoscale models use the hydrostatic approximation, once the grid resolution is small enough to resolve features whose horizontal width is of the same order as their height, a nonhydrostatic model must be used.

Nonhydrostatic models typically define the total pressure and density in terms of a *base state* or *reference state* that is in hydrostatic balance, and a nonhydrostatic perturbation. The base state varies only in height and is denoted with a subscript '0', while the perturbation varies in all three spatial dimensions as well as with time and is denoted with a '~'. The total pressure and density are then given by

$$p(x,y,z,t) = p_0(z) + \tilde{p}(x,y,z,t) \tag{13.9}$$

and

$$\rho(x,y,z,t) = \rho_0(z) + \tilde{\rho}(x,y,z,t). \tag{13.10}$$

The governing equations written in terms of (13.9) and (13.10) are

$$\frac{Du}{Dt} = -\frac{1}{\rho_0}\frac{\partial \tilde{p}}{\partial x} + 2\Omega \sin \phi\, v - 2\Omega \cos \phi\, w + F_x \tag{13.11}$$

$$\frac{Dv}{Dt} = -\frac{1}{\rho_0}\frac{\partial \tilde{p}}{\partial y} - 2\Omega \sin \phi\, u + F_y \tag{13.12}$$

$$\frac{Dw}{Dt} = -\frac{1}{\rho_0}\frac{\partial \tilde{p}}{\partial z} - 2\Omega \cos \phi\, u - \frac{\tilde{\rho}}{\rho_0}g + F_z \tag{13.13}$$

$$\frac{\partial \tilde{\rho}}{\partial t} + \nabla\left(\rho_0 \vec{V}\right) + \nabla\left(\tilde{\rho}\vec{V}\right) = 0 \tag{13.14}$$

$$c_p\frac{DT}{Dt} - \frac{1}{\rho_0}\frac{Dp_0}{Dt} - \frac{1}{\rho_0}\frac{D\tilde{p}}{Dt} = \dot{Q} + F_T \tag{13.15}$$

$$p = \rho R_d T\left(1 + 0.61q\right) \tag{13.16}$$

$$\frac{\partial(\tilde{\rho}q)}{\partial t} + \nabla \cdot \left(\rho_0 q\vec{V}\right) + \nabla \cdot \left(\tilde{\rho}q\vec{V}\right) = S_O - S_K + F_q, \tag{13.17}$$

The specific method for diagnosing the vertical velocity will depend on the vertical coordinate system used (pressure, height, sigma) and other factors specific to the particular model implementation. Details of how to diagnose the vertical velocity are not discussed in this book.

which along with (13.9) and (13.10) form a closed set. These are similar to the *Boussinesq* form of the equations, although we allow the base state density to change in the vertical, instead of keeping it constant with height as in the true Boussinesq equations.

Since the horizontal grid spacing is of the order of the vertical grid spacing, the restriction on the time interval due to the propagation of acoustic waves is the same in both the vertical and horizontal directions. Unless another means can be found for eliminating fast-moving acoustic waves from the model equations, nonhydrostatic models require a fairly short time interval and therefore more iterations are required to advance the model solution by a specified amount of time. This provides a segue into our discussion of the continuity equation in the next section.

13.4 The Continuity Equation

For nonhydrostatic models the equations are often converted to sigma coordinates; however, for purposes of illustration it will be easier for us to avoid that complication and leave them in Cartesian height coordinates.

13.4.1 Fully compressible models

The implementation of the equations for a fully compressible, nonhydrostatic model is fairly straightforward. Equations (13.11) through (13.14) are used to predict the new values of u, v, w, and $\tilde{\rho}$. Equation (13.15) is a prognostic equation for both T and $\tilde{p}$, but cannot be used for both. However, by expanding Dp_0/Dt as[2]

$$\frac{Dp_0}{Dt} = \frac{\partial p_0}{\partial t} + \vec{V} \cdot \nabla p_0 = w\frac{\partial p_0}{\partial z} = -w\rho_0 g, \qquad (13.18)$$

(13.15) becomes

$$c_p\frac{DT}{Dt} + gw - \frac{1}{\rho_0}\frac{D\tilde{p}}{Dt} = \dot{Q} + F_T. \qquad (13.19)$$

The term $\rho_0^{-1}D\tilde{p}/Dt$ is much smaller in magnitude than gw, as can be shown using scale analysis (see Exercise 13.1), so we can write

---

[2]Equation (13.18) is valid regardless of whether we use Cartesian or spherical coordinates.

(13.19) as

$$c_p \frac{DT}{Dt} + gw = \dot{Q} + F_T, \tag{13.20}$$

and use this for our predictive equation for temperature. Equation (13.17) is used to predict the humidity. Once this, along with u, v, w, $\tilde{\rho}$, and T, are predicted, the perturbation pressure is then diagnosed from the equation of state, (13.16).

13.4.2   Anelastic models

The fully compressible nonhydrostatic equations allow very fast acoustic waves as solutions and therefore require that a shorter time interval be used. By replacing the fully compressible continuity equation with the anelastic continuity equation,

$$\nabla \cdot \left(\rho_0 \vec{V} \right) = 0, \tag{13.21}$$

all acoustic waves are eliminated. However, by doing so we lose a predictive equation for density, and therefore we cannot diagnose the pressure. This makes the implementation of these equations in a model actually more difficult than for the fully compressible model.

    We can still predict u, v, w, and T as before but must find another route to diagnosing the pressure. We start by differentiating (13.21) with respect to time,

$$\frac{\partial}{\partial t} \left[\nabla \cdot \left(\rho_0 \vec{V} \right) \right] = 0. \tag{13.22}$$

Since partial derivatives commute, this is equivalent to

$$\nabla \cdot \left(\rho_0 \frac{\partial \vec{V}}{\partial t} \right) = 0. \tag{13.23}$$

The three momentum equations, (13.1), (13.2), and (13.3) are written in vector form as

$$\frac{\partial \vec{V}}{\partial t} = -\vec{V} \cdot \nabla \vec{V} - \frac{1}{\rho_0} \nabla \tilde{p} - 2\vec{\Omega} \times \vec{V} - \frac{\tilde{\rho}}{\rho_0} g\hat{k} + \vec{F}, \tag{13.24}$$

which on substitution into (13.23) results in

$$\nabla \cdot \left(\rho_0 \vec{V} \cdot \nabla \vec{V} + \nabla \tilde{p} + 2\rho_0 \vec{\Omega} \times \vec{V} + \hat{k} \tilde{\rho} g - \rho_0 \vec{F} \right) = 0. \tag{13.25}$$

Solving (13.25) for $\tilde{p}$ gives

$$\nabla^2 \tilde{p} = -\rho_0 \nabla \cdot \left(\vec{V} \cdot \nabla \vec{V} \right) - 2\rho_0 \nabla \cdot \left(\vec{\Omega} \times \vec{V} \right)$$
$$- g\frac{\partial \tilde{\rho}}{\partial z} + \rho_0 \nabla \cdot \vec{F}. \quad (13.26)$$

Equation (13.26) is an elliptic equation, and is computationally intensive to solve numerically. Thus, though use of the anelastic continuity equation allows use of a longer time interval and thus fewer time iterations to complete the simulation, this advantage may be offset by the fact that we now must solve an elliptic equation at every time step in order to diagnose the pressure. The anelastic continuity equation is sometimes used in 2D nonhydrostatic models, where the solution of the elliptic equation is not as computationally intensive as in 3D models. Most modern 3D nonhydrostatic models use the fully compressible continuity equation.

Exercises

Ex. 13.1: In the vicinity of convection typical values of u, v, and w are on the order of 10 m s^{-1}. Typical values for horizontal perturbation pressure gradient are 10^{-3} Pa m^{-1} and for vertical perturbation pressure gradient, 10^{-1} Pa m^{-1}. The perturbation pressure tendency may be of the order of a few millibars per hour. Show that, with these values, $\rho_0^{-1} D\tilde{p}/Dt$ is at least an order of magnitude smaller than gw.

## CHAPTER 14

Semi-implicit and Semi-Lagrangian Methods

14.1 Overcoming the CFL Stability Condition

In Chapter 4 we established that for most time-differencing schemes the largest time interval that may be used is directly proportional to the smallest grid spacing in the model. The stability condition has the form

$$|c\Delta t/d| \leq \alpha \tag{14.1}$$

where c is the speed of the fastest wave supported by the model equations, Δt is the time interval, and d is the smallest grid spacing. The value for α depends on the specific finite-differencing scheme being used and in general is a value of one or less. The exceptions to (14.1) are implicit schemes which allow arbitrarily long time increments to be used and still maintain stability.

The limitation on time increment imposed by (14.1) becomes especially severe in those applications requiring very small grid spacing. These include mesoscale and cloud-scale models, which require small grid spacing in order to resolve small features, and also global models using spherical coordinates, since the longitude lines become very close together in the polar regions and result in small grid spacing.

Two methods have been developed in attempts to loosen the restriction imposed on the maximum time increment by (14.1). These are *semi-implicit* time schemes and *semi-Lagrangian* methods, which are the subjects of this chapter. Our treatment will shy more toward the descriptive and qualitative, rather than quantitative. The aim is to familiarize the reader with the basic concepts of these techniques in order to make the advanced literature more accessible.

14.2   Semi-implicit Time-differencing

Fully implicit time schemes have the advantage of allowing an arbitrarily long time interval to be used, but are computationally intensive. A semi-implicit scheme is a compromise between a fully-explicit scheme and a fully implicit scheme, and allows a longer time step to be used while maintaining numerical stability.

14.2.1   Semi-implicit schemes defined

Semi-implicit schemes treat certain variables implicitly while treating others within the same equation explicitly. For example, consider a partial differential equation of the form

$$\frac{\partial u}{\partial t} = F(u, v, \eta).$$
(14.2)

We could employ a semi-implicit time-differencing scheme that uses information at time level n for the two variables u and v, but at time level $n+1$ for variable η. Such a scheme would have the form

$$u^{n+1} = u^n + \Delta t F(u^n, v^n, \eta^{n+1}),$$
(14.3)

in which u and v are treated explicitly while η is treated implicitly. Some schemes may even treat the same variable both explicitly and implicitly within the same equation, such as[1]

$$u^{n+1} = u^n + \Delta t F(u^n, v^n, \eta^n, \eta^{n+1}).$$
(14.4)

---

[1]If the forcing term is a function of only a single variable and the scheme uses values of that variable u at both $n + 1$ and n, such as $u^{n+1} = u^n + \Delta t F(u^n, u^{n+1})$, some authors would refer to this as a semi-implicit scheme, while others (including us) refer to it as a fully implicit scheme. As an example of such a scheme see the trapezoidal scheme in Appendix B.

14.2.2 Application to the primitive equations

In the stability analysis of the barotropic primitive equations in Chapter 8 we concluded that the maximum time interval is dictated by the speed of the fast-moving gravity waves. The terms in the governing equations that control this wave speed are the pressure gradient terms and the divergence terms. The Coriolis terms do not appreciably impact the speed of the waves and therefore do not impact the stability of the scheme. We apply a *semi-implicit* scheme to these equations by treating the pressure gradient and divergence terms implicitly, while explicitly treating the Coriolis terms.

This method can be used for either the barotropic or full primitive equations. For simplicity we illustrate it using the nonlinear barotropic primitive equations

$$\frac{\partial u}{\partial t} + u\frac{\partial u}{\partial x} + v\frac{\partial u}{\partial y} = -g'\frac{\partial \eta}{\partial x} + fv \tag{14.5}$$

$$\frac{\partial v}{\partial t} + u\frac{\partial v}{\partial x} + v\frac{\partial v}{\partial y} = -g'\frac{\partial \eta}{\partial y} - fu \tag{14.6}$$

$$\frac{\partial \eta}{\partial t} + u\frac{\partial \eta}{\partial x} + v\frac{\partial \eta}{\partial y} = -H\left(\frac{\partial u}{\partial x} + \frac{\partial v}{\partial y}\right) - \eta\left(\frac{\partial u}{\partial x} + \frac{\partial v}{\partial y}\right). \tag{14.7}$$

It is the pressure gradient terms $g'\partial\eta/\partial x$, $g'\partial\eta/\partial y$, and the linear divergence term $H\left(\partial u/\partial x + \partial v/\partial y\right)$, that are the dominant terms in dictating the speed of the gravity waves so it is these terms that we will treat implicitly. All remaining terms are treated explicitly.

For illustration we write the time derivatives using finite differences, but leave the spatial derivatives as they are, since the specific differencing scheme used for the spatial derivatives is not important for understanding the semi-implicit method. The finite-difference forms of (14.5) through (14.7) are then

$$\frac{u^{n+1} - u^{n-1}}{2\Delta t} + u^n\frac{\partial u^n}{\partial x} + v^n\frac{\partial u^n}{\partial y}$$

$$= -\frac{g'}{2}\left(\frac{\partial \eta^{n+1}}{\partial x} + \frac{\partial \eta^{n-1}}{\partial x}\right) + fv^n \tag{14.8}$$

$$\frac{v^{n+1} - v^{n-1}}{2\Delta t} + u^n \frac{\partial v^n}{\partial x} + v^n \frac{\partial v^n}{\partial y}$$

$$= -\frac{g'}{2} \left(\frac{\partial \eta^{n+1}}{\partial y} + \frac{\partial \eta^{n-1}}{\partial y} \right) - f u^n \quad (14.9)$$

$$\frac{\eta^{n+1} - \eta^{n-1}}{2\Delta t} + u^n \frac{\partial \eta^n}{\partial x} + v^n \frac{\partial \eta^n}{\partial y}$$

$$= -\frac{H}{2} \left(\frac{\partial u^{n+1}}{\partial x} + \frac{\partial v^{n+1}}{\partial y} + \frac{\partial u^{n-1}}{\partial x} + \frac{\partial v^{n-1}}{\partial y} \right)$$

$$- \eta^n \left(\frac{\partial u^n}{\partial x} + \frac{\partial v^n}{\partial y} \right). \quad (14.10)$$

Notice that for the pressure gradient and linear divergence terms we are averaging the spatial derivative between times $n-1$ and $n+1$.

The next step is to combine (14.8) through (14.10) into a single equation in η^{n+1}, eliminating u^{n+1} and v^{n+1}. This is achieved by writing (14.8) and (14.9) each in the form of $u^{n+1} = \ldots$ and $v^{n+1} = \ldots$ and then substituting the right-hand side of these expressions into (14.10). The resulting, very complicated equation is

$$\nabla^2 \eta^{n+1} - \frac{\eta^{n+1}}{g'H\Delta t^2} =$$

$$- \nabla^2 \eta^{n-1} - \frac{\eta^{n-1}}{g'H\Delta t^2} + \frac{1}{g'\Delta t} \left(\frac{\partial u^{n-1}}{\partial x} + \frac{\partial v^{n-1}}{\partial y} \right)$$

$$- \frac{2}{g'} \left[\frac{\partial}{\partial x} \left(u^n \frac{\partial u^n}{\partial x} + v^n \frac{\partial u^n}{\partial y} \right) + \frac{\partial}{\partial y} \left(u^n \frac{\partial v^n}{\partial x} + v^n \frac{\partial v^n}{\partial y} \right) \right]$$

$$+ \frac{2}{g'H\Delta t} \left(u^n \frac{\partial \eta^n}{\partial x} + v^n \frac{\partial \eta^n}{\partial y} \right) + \frac{2f}{g'} \left(\frac{\partial v^n}{\partial x} - \frac{\partial u^n}{\partial y} \right)$$

$$+ \frac{2\eta^n}{g'H\Delta t} \left(\frac{\partial u^n}{\partial x} + \frac{\partial v^n}{\partial y} \right). \quad (14.11)$$

This is a nonhomogeneous elliptic equation in η^{n+1}, with all values on the right-hand side known. Equation (14.11) is solved for η^{n+1}, and this value is then used in (14.8) and (14.9) to obtain the new values for u^{n+1} and v^{n+1}.

Even though the semi-implicit method requires the solution of a computationally intensive elliptic equation, the longer time interval allows fewer iterations to advance the model simulation by a specified time period. However, one major drawback of the semi-implicit scheme is that its wave dispersion properties are such that gravity waves propagate much slower than they should. For larger-scale model simulations where Rossby waves are the most relevant dynamical feature, this slowing of the gravity waves may not be an issue and a semi-implicit scheme performs very well. But for scales and phenomena where the geostrophic adjustment process due to gravity wave propagation is important, the semi-implicit scheme may not be appropriate.

We illustrated the semi-implicit method for the barotropic primitive equations. Its application to the full primitive equations, not shown here, would proceed along a similar outline.

14.3 Semi-Lagrangian Methods

14.3.1 The semi-Lagrangian method described

So far we have been dealing with prognostic equations in their Eulerian form, meaning that we have been trying to predict local tendencies at a fixed point in space. This is indicated by the fact that all our predictive equations for an arbitrary scalar s have been written in terms of the partial derivative with respect to time,

$$\frac{\partial s}{\partial t} = \cdots . \tag{14.12}$$

Another approach is to estimate the time derivative following an individual air parcel, using the material derivative and equations of the form

$$\frac{Ds}{Dt} = \cdots . \tag{14.13}$$

Equations written in the form of (14.12) are called *Eulerian*, while those in the form (14.13) are called *Lagrangian*. We can always switch between the Eulerian and Lagrangian forms using the identity

$$\frac{Ds}{Dt} = \frac{\partial s}{\partial t} + \vec{V} \cdot \nabla s. \tag{14.14}$$

In a *Lagrangian model* we would approximate (14.13) using finite differences to obtain a predictive equation for changes following a fluid parcel. We would start out with a population of air parcels at the beginning of the model run and track their trajectories and the changes to them as they traverse the model domain. The problem with this approach is that the fluid parcels will not remain evenly distributed within the domain, but will instead congregate in regions of convergence and leave large areas of the domain void of parcels in regions of divergence.

A compromise between the Lagrangian and Eulerian approaches is the *semi-Lagrangian* approach. Instead of tracking individual air parcels as they move in the domain, we continue to define the model variables at individual model grid points. We then predict the changes to the variables at the grid points by determining where the air parcels that affect a particular model grid point at time level n were located at the previous time level, $n - 1$, and then predicting the changes to these parcels while they are enroute to our model grid points.

There are three key steps to the semi-Lagrangian approach. They are:

1. *Calculating back trajectories* from each three-dimensional grid point, P, to find the point $\tilde{P}$ at which the air parcel was located at the previous time level $n - 1$. Representing the coordinates of the grid point P by (x_p, y_p, z_p) and those of point $\tilde{P}$ by $(\tilde{x}_p, \tilde{y}_p, \tilde{z}_p)$, the origin of the air parcel is found by numerically integrating the equations

$$\tilde{x}_P = x_P - \int_{(n-1)\Delta t}^{n\Delta t} u \, dt \qquad (14.15)$$

$$\tilde{y}_P = y_P - \int_{(n-1)\Delta t}^{n\Delta t} v \, dt \qquad (14.16)$$

$$\tilde{z}_P = z_P - \int_{(n-1)\Delta t}^{n\Delta t} w \, dt. \qquad (14.17)$$

2. *Calculating the properties the air parcel* had at time level $n-1$, when it was at coordinates $\tilde{x}_P$, $\tilde{y}_P$, and $\tilde{z}_P$.

3. *Calculating the changes to the air parcel's properties* as it moves along the trajectory between the point $\tilde{P}$ and point P.

14.3.2   Advantages and disadvantages

There are several difficulties with implementing the semi-Lagrangian approach:

- It is unlikely that the back trajectory calculations will place the air parcels exactly at a model grid point at the previous time step. Thus, some sort of interpolation must be used to determine the properties of the air parcel in terms of the surrounding grid points. These can range from a simple, linear approximation to complex spline methods.

- The winds advecting the air parcels are not likely to be constant along the trajectory.

- Unless the properties of the air parcel are considered to be conserved along its path, then changes to the properties enroute must be taken into account.

- Finite-difference schemes applied to the semi-Lagrangian method often do not conserve quantities that should be conserved, requiring a periodic renormalization to avoid accumulations or losses of conserved properties.

Despite these difficulties, semi-Lagrangian techniques have several advantages. One advantage is that longer time steps can often be used without the simulation becoming unstable. One interpretation of the traditional CFL stability criteria for Eulerian models, $c\Delta t/d \leq 1$, is that an air parcel should not move more than one grid cell during a single time step or aliasing errors may occur. But with semi-Lagrangian methods we are tracking an air parcel from its point of origin and there is no problem with it passing through multiple grid cells during a single time step. The longer time step allowed by the semi-Lagrangian method can often offset the extra computational load of having to calculate the back trajectories. Another advantage of semi-Lagrangian methods is that they are well suited to variable grid spacings and irregular-shaped grids.

Semi-Lagrangian techniques in combination with semi-implicit time differencing are now widely used in atmospheric numerical modeling, particularly in large-scale global models.

## CHAPTER 15

Finite-volume Methods

15.1 Flux-form Equations

15.1.1 Flux-form equations and conservation laws

Many atmospheric processes are governed by conservation laws of the form

$$\frac{\partial s}{\partial t} + \nabla \cdot (s\vec{V}) = G, \tag{15.1}$$

where s is the amount (usually mass or moles) per unit volume of a substance, and G represents sources and sinks of the substance.[1] The conserved substance can be any atmospheric constituent or trace gas. For example, if the conserved substance is air, then s is simply the air density ρ, and G is zero, in which case (15.1) becomes the mass continuity equation. If the conserved substance is water vapor, then s is the absolute humidity ρ_v, and G represents sources and sinks of water vapor such as evaporation and condensation. The product $s\vec{V}$ has units of amount (e.g., mass, moles, or number) per area per time, and is the flux of the substance. The term $\nabla \cdot (s\vec{V})$ is the flux divergence. In the absence of sources, sinks, or turbulent transport, (15.1) has the physical interpretation that the change in s

---

[1] If s and $\vec{V}$ are Reynolds-averaged quantities, then G also includes the turbulent transport of s.

at a fixed point in space is due solely to flux divergence at the point. Through the vector identity $\nabla \cdot (s\vec{V}) = \vec{V} \cdot \nabla s + s\nabla \cdot \vec{V}$, conservation equations such as (15.1) can also be written as

$$\frac{\partial s}{\partial t} + \vec{V} \cdot \nabla s + s\nabla \cdot \vec{V} = G. \tag{15.2}$$

In the absence of sources, sinks, or turbulent transport, (15.2) has the physical interpretation that s at a fixed point changes only in response to advection or to divergence of the wind field. Equation (15.1) is the *flux-form* equation for s, while (15.2) is the *advective-form*.

15.1.2   Advantage of flux-form equations

If we were creating a model to predict changes in s with time we could choose either the flux-form or the advective form of the equation as the basis for the model. However, the flux-form equation has advantages over the advective form. We illustrate this using a one-dimensional example without sources, sinks, or turbulent transport, so that the flux-form and advective-form equations are

$$\frac{\partial s}{\partial t} + \frac{\partial (su)}{\partial x} = 0 \tag{15.3}$$

and

$$\frac{\partial s}{\partial t} + u\frac{\partial s}{\partial x} + s\frac{\partial u}{\partial x} = 0 \tag{15.4}$$

respectively. Using centered differencing the finite-difference form of (15.3) is

$$s_i^{n+1} = s_i^{n-1} - \frac{\Delta t}{\Delta x}\left[(su)_{i+1}^n - (su)_{i-1}^n\right], \tag{15.5}$$

while the finite-difference form of (15.4) is

$$s_i^{n+1} = s_i^{n-1} - \frac{\Delta t}{\Delta x}u_i^n\left[s_{i+1}^n - s_{i-1}^n\right] - \frac{\Delta t}{\Delta x}s_i^n\left(u_{i+1}^n - u_{i-1}^n\right). \tag{15.6}$$

Either (15.5) or (15.6) could be used in our model, but the flux-form version, (15.5), contains fewer terms, is simpler to implement, and also does a better job of conserving the amount of the substance s in the model.

15.1.3 The momentum equation in flux form

The vector-form of the momentum equation is

$$\frac{\partial \vec{V}}{\partial t} + \vec{V} \cdot \nabla \vec{V} = -\frac{1}{\rho} \nabla p - 2\vec{\Omega} \times \vec{V} + \vec{g} + \vec{F} \qquad (15.7)$$

where $\vec{F}$ is a vector representing the effects of turbulent momentum transport. The momentum equation can also be expressed in flux form as

$$\frac{\partial (\rho \vec{V})}{\partial t} + \nabla \cdot (\rho \vec{V} \vec{V}) = -\nabla p - 2\vec{\Omega} \times (\rho \vec{V}) + \rho \vec{g} + \rho \vec{F} . \qquad (15.8)$$

The term $\rho \vec{V}$ is the *momentum per unit volume*, while $\rho \vec{V} \vec{V}$ is the *momentum flux*.[2] It is notable that if we define the forcing terms on the right-hand side of (15.8) by $\vec{G}$, and the momentum per unit volume as $\vec{S}$, then (15.8) becomes

$$\frac{\partial \vec{S}}{\partial t} + \nabla \cdot (\vec{S} \vec{V}) = \vec{G}, \qquad (15.9)$$

which has the same form as (15.1). If there are no forcing terms, then the physical interpretation of (15.8) or (15.9) is that the momentum per unit volume at a fixed point can change only due to divergence or convergence of momentum flux. In Cartesian coordinates the component equations of (15.8) are

$$\frac{\partial \rho u}{\partial t} + \frac{\partial (\rho u u)}{\partial x} + \frac{\partial (\rho u v)}{\partial y} + \frac{\partial (\rho u w)}{\partial z}$$
$$= -\frac{\partial p}{\partial x} + 2\Omega \rho v \sin \phi - 2\Omega \rho w \cos \phi + F_x, \quad (15.10)$$

$$\frac{\partial \rho v}{\partial t} + \frac{\partial (\rho v u)}{\partial x} + \frac{\partial (\rho v v)}{\partial y} + \frac{\partial (\rho v w)}{\partial z}$$
$$= -\frac{\partial p}{\partial y} - 2\Omega \rho u \sin \phi + F_y, \quad (15.11)$$

---

[2]The product, $\vec{V} \vec{V}$, of two vectors is a perfectly legal mathematical operation, and is called the *tensor product* because it results in a second-order tensor. The momentum flux, $\rho \vec{V} \vec{V}$, is therefore a second-order tensor.

and

$$\frac{\partial \rho w}{\partial t} + \frac{\partial \left(\rho w u\right)}{\partial x} + \frac{\partial \left(\rho w v\right)}{\partial y} + \frac{\partial \left(\rho w w\right)}{\partial z}$$

$$= -\frac{\partial p}{\partial z} + 2\Omega \rho u \cos \phi - \rho g + F_z. \quad (15.12)$$

Finite-difference approximations of the flux-form of the momentum equations do a better job of conserving momentum than do those for the advective-form of the momentum equations. For this reason, many primitive equation models use the flux-form momentum equations.

15.2   Finite-volume Versus Finite-difference

In our previous discussions on modeling we have been assuming that the values of the variables at a model grid point are only valid at the grid point itself, and tell us nothing about the distributions of the variables between the grid points. We have made no assumptions about the distribution of the variables between the grid points except to perhaps pretend that they vary linearly between grid points. Another approach, called the *finite-volume method*, assumes that the values at the grid points are *volume-averaged* values of the variables over the three-dimensional *grid cell* surrounding the grid point. In order to implement a finite-volume method we are forced to make assumptions about the distribution of the variables between grid points. If we make the simplest assumption of linearity, then the finite-volume method is identical to the finite-difference approach. More complex assumptions regarding the distribution of variables between grid points render the finite-volume different (and hopefully more accurate) than the finite-difference approach.

15.3   The Finite-volume Method

15.3.1   The divergence theorem

The mathematics behind the finite-volume method is based on the *divergence theorem*,[3] which states that the volume integral of the divergence of a vector is equal to the surface integral of the normal

---

[3] Also known as Gauss' theorem.

component of the vector. In general form this is expressed mathematically as

$$\iiint \nabla \cdot \vec{B} d\Lambda = \iint \vec{B} \cdot \hat{n} dA \tag{15.13}$$

where Λ indicates volume and A indicates area. The vector $\hat{n}$ is a unit vector perpendicular to the surface dA and points outward from the interior of the volume. Starting with a conservation law of the form (15.1) we integrate it over a volume, and then divide by the volume to get

$$\frac{1}{\Lambda} \iiint \frac{\partial s}{\partial t} d\Lambda + \frac{1}{\Lambda} \iiint \nabla \cdot (s\vec{V}) d\Lambda = \frac{1}{\Lambda} \iiint G d\Lambda. \tag{15.14}$$

The first term of (15.14) is

$$\frac{1}{\Lambda} \iiint \frac{\partial s}{\partial t} d\Lambda = \frac{\partial}{\partial t} \left(\frac{1}{\Lambda} \iiint s d\Lambda \right) = \frac{\partial \tilde{s}}{\partial t}, \tag{15.15}$$

where $\tilde{s}$ indicates a volume-averaged value of s.[4] Likewise, the last term in (15.14) is just the volume-averaged sources and sinks term, $\tilde{G}$. Applying the divergence theorem to the middle term of (15.14) yields

$$\frac{1}{\Lambda} \iiint \nabla \cdot \left(s\vec{V} \right) d\Lambda = \frac{1}{\Lambda} \iint s\vec{V} \cdot \hat{n} dA, \tag{15.16}$$

which is the surface-integrated flux divided by the volume. Equation (15.14) has now become

$$\frac{\partial \tilde{s}}{\partial t} + \frac{1}{\Lambda} \iint s\vec{V} \cdot \hat{n} dA = \tilde{G}. \tag{15.17}$$

Equation (15.17) is a prognostic equation for the volume-averaged value $\tilde{s}$, as compared to (15.1), which is a prognostic equation for the value of s at a particular point in space.

15.3.2 Application of the finite-volume method

To illustrate how the finite-volume method is applied we use the one-dimensional equivalent of (15.17) with no source or sink terms,

---

[4]We use a tilde rather than an overbar to indicate the volume average, since we have previously used the overbar to indicate a Reynolds-averaged quantity.

and centered-in-time differencing. The model equation would then be

$$\frac{\tilde{s}_i^{n+1} - \tilde{s}_i^{n-1}}{2\Delta t} + \frac{1}{\Delta x \Delta y \Delta z} \left[\langle su \rangle_{i+1/2}^n - \langle su \rangle_{i-1/2}^n \right] \Delta y \Delta z = 0 \quad (15.18)$$

or

$$\tilde{s}_i^{n+1} = \tilde{s}_i^{n-1} - \frac{2\Delta t}{\Delta x} \left[\langle su \rangle_{i+1/2}^n - \langle su \rangle_{i-1/2}^n \right]. \quad (15.19)$$

The $\langle su \rangle$ terms appearing in (15.18) and (15.19) represent the time-averaged fluxes at the boundaries of the grid volume,

$$\langle su \rangle = \frac{1}{\Delta t} \int_{n\Delta t}^{(n+1)\Delta t} su \, dt. \quad (15.20)$$

These fluxes are averaged over the time interval Δt, and are valid halfway between the model grid points, which is why they are indicated with fractional grid indices. In order to implement (15.19) an assumption must be made as to how these time-averaged fluxes are computed. In the very simplest case we just average the value su of the adjacent grid volumes and use this for the time-averaged flux between the two grid volumes,[5]

$$\langle su \rangle_{i+1/2}^n = \frac{(\tilde{s}\tilde{u})_{i+1}^n + (\tilde{s}\tilde{u})_i^n}{2} \quad (15.21)$$

and

$$(su)_{i-1/2}^n = \frac{(\tilde{s}\tilde{u})_i^n + (\tilde{s}\tilde{u})_{i-1}^n}{2}. \quad (15.22)$$

Then (15.19) becomes

$$\tilde{s}_i^{n+1} = \tilde{s}_i^{n-1} - \frac{\Delta t}{\Delta x} \left[(\tilde{s}\tilde{u})_{i+1}^n - (\tilde{s}\tilde{u})_{i-1}^n \right] \quad (15.23)$$

which is identical in form to (15.5). The only difference between (15.5) and (15.23) is that, in the former, the variables are valid only at a grid point, while in the latter the variables are averaged over a grid cell. However, if more complex assumptions are made regarding the distribution of variables between grid points, then the finite-difference and finite-volume schemes will have different forms.

---

[5]This is equivalent to assuming a linear fit between grid points.

15.3.3 Advantage of the finite-volume method

For variables that have smooth distributions across a model grid, finite-difference and finite-volume techniques will produce similar results. For variables that are discontinuous, the finite-volume method will be superior to finite-difference methods. Thus, for modeling of chemical plumes, or bursts of passive tracers, a finite-volume scheme would be preferred. Another advantage of finite-volume schemes is that they are readily adaptable to irregular spaced (nonuniform) grids. There is also no requirement that the grid volume be cube-like. Any closed 3D shape can be used to define the volume. Tetrahedral (four-sided) grid cells can be used and have the advantage of only requiring calculations of fluxes at four faces, rather than at six faces for a cubic grid cell.

15.4 Flux-form Semi-Lagrangian (FFSL) Method

The main difference between the various finite-volume methods lies in how the time-averaged fluxes at the volume edges are determined. There are many different possible methods for approximating the integral on the right-hand side of (15.20). Of particular note are the *flux-form semi-Lagrangian or FFSL* transport schemes, which have found widespread use in chemical transport models. There are several different specific schemes that fall into this category, but they all share the characteristic that they are *upstream-biased*, meaning that the time-averaged fluxes are determined using only information upstream of the grid volume boundary.

An example of an FFSL scheme is the *Van Leer* scheme, which has many desirable features. The scheme is *positive definite*, meaning that it will never result in negative values for the quantity being transported. This is especially important since negative values of atmospheric constituents are unphysical. This scheme will also not create any new local maxima or minima in the quantity being transported.

We will illustrate the one-dimensional version of the Van Leer scheme.[6] We use a staggered grid and indicate the indices of the

---

[6]The fact that we are illustrating the Van Leer scheme does not imply that it is the best of the FFSL schemes. We use it merely because it is one of the least complex FFSL schemes. Van Leer type schemes have, however, found widespread

u-grid volumes with capital letters, while using lower case letters for the s-grid indices. The u-grid volume with index $I = 0$ lies to the left of the s-grid volume with index $i = 0$. Note that in this configuration the centers of the grid volumes containing the u-values coincide with the edges of the grid volumes containing the values of s. On this grid the finite-difference form of (15.3), using forward differencing, is

$$\tilde{s}_i^{n+1} = \tilde{s}_i^n - \frac{\Delta t}{d} \left[\langle su \rangle_{I+1}^n - \langle su \rangle_I^n \right]. \tag{15.24}$$

The fluxes are then approximated using (15.25), the form used depending on whether $\tilde{u}_I^n$ is positive or negative,

$$\langle su \rangle_I^n = \begin{cases} \tilde{u}_I^n [\tilde{s}_{i-1}^n + (1/2)(1 - \tilde{u}_I^n \Delta t/d) \delta_{i-1}], & \text{for } \tilde{u}_I^n \geq 0; \\ \tilde{u}_I^n [\tilde{s}_i^n - (1/2)(1 - \tilde{u}_I^n \Delta t/d) \delta_i], & \text{for } \tilde{u}_I^n < 0. \end{cases} \tag{15.25}$$

The value of δ_i depends on whether the quantity

$$\sigma_i = \left(\tilde{s}_i^n - \tilde{s}_{i-1}^n \right) \left(\tilde{s}_{i+1}^n - \tilde{s}_i^n \right) \tag{15.26}$$

is positive or negative, and is found from

$$\delta_i = \begin{cases} 2\sigma_i / \left(\tilde{s}_{i+1}^n - \tilde{s}_{i-1}^n \right), & \text{for } \sigma_i \geq 0; \\ 0, & \text{for } \sigma_i < 0. \end{cases} \tag{15.27}$$

15.5 Extension to Multiple Dimensions

Extending flux-form schemes to multiple dimensional problems poses additional difficulties. This is because in one-dimensional models the fluxes are always parallel to the wind vector, while in two or three dimensions the fluxes through a particular grid-cell face may actually be at an oblique angle to the wind vector. This requires more complex algorithms, which are beyond the scope of this book, to correct for these transverse fluxes. Another difficulty arises because the complexity of the advection schemes may make it necessary to employ *operator splitting*, applying the flux calculation

use in atmospheric chemical transport modeling.

for the different coordinate axes sequentially, rather than simultaneously. Operator splitting introduces additional errors for which corrections must be applied. Appendix D contains more details on operator splitting.

CHAPTER 16

Chemical Transport Models

16.1 Introduction

One important application of numerical modeling is forecasting the transport and concentrations of trace gases and aerosols. Such models are called *chemical transport models*, often abbreviated as *CTM*. Chemical transport models are used for both research and operational purposes.

16.1.1 Air quality models

Air quality models are used for estimating and forecasting the amount of pollution in the air. These models are often employed by local and regional governments to monitor compliance with National Ambient Air Quality Standards (NAAQS), and for developing regulatory strategies for becoming compliant. These models can be used to perform simulated checks on the operation of proposed industrial facilities to ensure they will be in compliance with the standards. The models are also used to provide industrial plant design information for effective control strategies to reduce emissions of harmful air pollutants. Once a new facility is in operation, the facility can be checked to see if it actually is in compliance; this acts as a check on the reliability of the models in such design studies.

16.1.2   Dispersion models

These models are used by the military and emergency management personnel in planning for the accidental or intentional release of chemicals, radioactive material, or other materials. The results of dispersion model experiments can provide an estimate of the location of the areas most likely to be impacted, and to what degree they could be impacted. These scenarios are used in designing training, and for determining protective actions appropriate in the event of an actual release. If a release occurs, these same models are used to provide emergency managers with toxic dispersions forecasts.

16.1.3   Research models

Many trace gases and aerosols impact the radiative balance of the atmosphere. These trace gases and aerosols may have both natural and anthropogenic sources. Since changes in the concentrations of these trace gases and aerosols can impact climate, understanding and modeling of these constituents is an important application of CTMs.

16.2   Formulation of a Chemical Transport Model

The basis of all chemical transport models is the conservation law for a chemical species or aerosol, which in flux-form is

$$\frac{\partial s}{\partial t} + \nabla \cdot (s\vec{V}) = G. \qquad (16.1)$$

In (16.1), s is the amount of substance per unit volume, with the amount expressed in mass, moles, or number of molecules. The term G represents any sources or sinks of the substance, and may also include transport due to turbulent eddies. We can expand the right-hand side of (16.1) as

$$\frac{\partial s}{\partial t} + \nabla \cdot (s\vec{V}) = G^+ - G^- + R + F_s, \qquad (16.2)$$

where G^+ represents sources from processes other than chemical reactions, G^- represents sinks from sources other than chemical reactions, R represents sources or sinks from chemical reactions, and F_s is the turbulent flux term. The turbulent flux term is also often called the *diffusion* term, since it represents diffusion due to turbulent eddies. It is often parameterized as

$$F_s = K\nabla^2 s, \tag{16.3}$$

where K is the eddy diffusivity.

16.2.1 Passive tracers

In many applications it is the transport of passive tracers that are of interest. A passive tracer does not react with other atmospheric constituents, so the R term in (16.2) would be zero. An example would be the simulation of a chemical or radioactive plume or dust cloud. In this application the simulation would begin with an initial distribution of the tracer, and then the tracer would simply be redistributed by the atmospheric flow, also keeping the source and sink terms (G^+ and G^-) in (16.2) as zero. Another application of a passive tracer model would be to simulate the plume from a smokestack, in which case the source term, G^+, would be nonzero. If the passive tracer were removed from the atmosphere via wet or dry deposition than the sink term, G^-, would also be nonzero.

16.2.2 Reactive tracers

More complex chemical transport models not only model the transport of chemical constituents, but also model the interactions between constituents due to chemical reactions. In these models there is an equation of the form (16.2) for every chemical constituent in the model. If there are M constituents, then there would be M equations of the form

$$\frac{\partial s_1}{\partial t} + \nabla \cdot (s_1 \vec{V}) = G_1^+ - G_1^- + R_1 + F_{s_1}$$

$$\frac{\partial s_2}{\partial t} + \nabla \cdot (s_2 \vec{V}) = G_2^+ - G_2^- + R_2 + F_{s_2}$$

$$\vdots$$

$$\frac{\partial s_m}{\partial t} + \nabla \cdot (s_m \vec{V}) = G_m^+ - G_m^- + R_m + F_{s_m} \tag{16.4}$$

$$\vdots$$

$$\frac{\partial s_M}{\partial t} + \nabla \cdot (s_M \vec{V}) = G_M^+ - G_M^- + R_M + F_{s_M}.$$

These equations are not all independent, but are coupled via the reaction terms, R_m. For example, if s_1, s_2, s_3 and s_4 represent the concentrations of NO, O_3, NO_2, and O_2, the equations for these four constituents would be coupled via the chemical reaction

$$NO + O_3 \rightarrow NO_2 + O_2. \tag{16.5}$$

16.3   Components of a Chemical Transport Model

The components of a chemical transport model are: 1) a *transport scheme* for redistributing the constituents based on the winds; 2) a *chemical solver* to calculate the production and loss of constituents due to chemical reactions; 3) a scheme for accounting for sources of constituents from processes other than chemical reactions with each other; and 4) a scheme for accounting for nonreactive sinks of constituents.

16.3.1   The transport scheme

The transport scheme calculates the redistribution of the constituents by the wind. It accounts for the effects of the $\nabla \cdot (s\vec{V})$ and F_s terms in (16.4). The transport scheme may either be written in finite-difference or in finite-volume form, with the latter being more common. There are two main methods for obtaining the wind fields to drive the transport scheme. In the *offline method* the winds are obtained from a completely separate meteorological model that has been previously run. The wind fields from this model are saved in certain increments (though not at every time level) and then read into the chemical transport model and used to drive the transport scheme. The alternative is the *online method*, in which the chemical transport model is embedded within the meteorological model and the winds at each time level are used to drive the chemical transport model. Although the online method is preferred, it is also very computationally intensive.

The transport scheme also must account for the turbulent flux terms, which are responsible for the turbulent diffusion of the atmospheric constituents. This process must be parameterized using wind and thermodynamic information from the meteorological model that drives the transport model.

16.3.2 Nonreactive sources

Sources of the chemical constituents from mechanisms other than chemical reactions include emission from both natural and anthropogenic sources. Examples include gases emitted from factory smokestacks, biomass burning, nitrogen oxides produced from lightning discharges, etc. These are usually calculated ahead of time using emission inventories. These sources can be one of the more uncertain aspects of chemical transport modeling. They can be highly variable both spatially and temporally.

16.3.3 Nonreactive sinks

The loss of chemical constituents from processes other than chemical reactions is primarily due to dry and wet deposition. These processes must be parameterized using wind and thermodynamic information from the meteorological model that drives the chemical transport model. Wet deposition in particular is dependent on the distribution of hydrometeors, and must use information from the parent meteorological model.

16.4 Chemical Solvers

Often the most computationally intensive part of a complex chemical transport model is the chemical solver, which must calculate the production and loss of each constituent due to chemical reactions with the other constituents. There are often dozens of such constituents that are included in chemical transport models. In this section we briefly describe some aspects of how these reactions are simulated.

16.4.1 Reaction rates and coefficients

For a dissociation reaction of the form

$$A \rightarrow B + C \tag{16.6}$$

the rate of change of the concentration of A is given by

$$\frac{d[A]}{dt} = -k_A [A] \tag{16.7}$$

where the brackets denote the concentration of species A (usually in molecules per volume). The coefficient k_A is called the *rate coefficient*.

The rates of change of species B and C are given by

$$\frac{d\,[\mathrm{B}]}{dt} = \frac{d\,[\mathrm{C}]}{dt} = -\frac{d\,[\mathrm{A}]}{dt} = k_A\,[\mathrm{A}]\,, \qquad (16.8)$$

since every unit of A that is lost produces one unit each of B and C.

For a reaction where two molecules react to form a single product, such as[1]

$$\mathrm{B + C + M} \;\rightarrow\; \mathrm{A + M} \qquad (16.9)$$

the rate of change equations are

$$\frac{d\,[\mathrm{B}]}{dt} = \frac{d\,[\mathrm{C}]}{dt} = -\frac{d\,[\mathrm{A}]}{dt} = -k_{\mathrm{BC}}\,[\mathrm{B}]\,[\mathrm{C}]\,. \qquad (16.10)$$

Some dissociation reactions require a photon of solar energy,

$$\mathrm{B} + h\nu \;\rightarrow\; \mathrm{D + E}, \qquad (16.11)$$

in which case the rate coefficient is called a *photolysis rate coefficient* denoted by J, and the rate equations are

$$\frac{d\,[\mathrm{D}]}{dt} = \frac{d\,[\mathrm{E}]}{dt} = -\frac{d\,[\mathrm{B}]}{dt} = J_{\mathrm{B}}\,[\mathrm{B}]\,. \qquad (16.12)$$

The rate coefficients are in general functions of both temperature and pressure. The photolysis rate coefficients are functions of the *actinic flux*, which depends on the sun angle, cloud cover, and cloud distribution.[2]

16.4.2　A coupled system of rate equations

Supposing that species A, B, C, D, and E were the only five species in our model, and that reactions (16.6), (16.9), and (16.11) were the only possible reactions between the species. We could then write

---

[1]In these reactions M represents a third, inert molecule that serves to carry away excess kinetic energy, leaving a stable product. Since the concentration of M does not change, we do not need to include a rate of change equation for M. In the atmosphere, M is most likely N_2 or O_2, since these are the two most abundant molecules.

[2]See Chapter 10, Section 10.4.7

the following ordinary differential equations for the concentrations of the five species:

$$\frac{d\,[A]}{dt} = -k_A\,[A] + k_{BC}\,[B]\,[C]$$

$$\frac{d\,[B]}{dt} = k_A\,[A] - k_{BC}\,[B]\,[C] - J_B\,[B]$$

$$\frac{d\,[C]}{dt} = -k_{BC}\,[B]\,[C] \tag{16.13}$$

$$\frac{d\,[D]}{dt} = J_B\,[B]$$

$$\frac{d\,[E]}{dt} = J_B\,[B]\,.$$

This is a coupled set of nonlinear ordinary differential equations, and they must be solved numerically in order to predict the changes in each chemical species due to the reactions between the species.

16.4.3 Solving the system of rate equations

Theoretically we could simply solve each of the equations in (16.13) using an explicit finite-difference scheme. In practice this is difficult, because the reaction rates may vary by orders of magnitude, and the time step is limited by the fastest reaction. Also, explicit schemes do not conserve the total mass of the chemical constituents. Explicit methods are therefore not practical for atmospheric chemical solvers. Simple implicit and semi-implicit methods allow a longer time step, but are more computationally intensive and also may not conserve mass. Our problems are further compounded by the fact that in the real atmosphere there are dozens of chemical constituents and hundreds or even thousands of possible reactions between them.

Chemical solvers therefore rely on very complex and highly developed methods for solving systems of equations such as (16.13). Some of the methods employed are the quasi-steady-state-approximation (QSSA), backward differentiation formulas (BDF), implicit Runge-Kutta schemes, Rosenbrock methods, and extrapolation. The details of these methods are beyond the scope of this book.[3]

---

[3]A good reference for delving more into the methods used in chemical solvers

is Chapter 12 of Jacobsen, M.Z, 2005: *Fundamentals of Atmospheric Modeling (2 ed.)*, Cambridge University Press, 828 pp.

## CHAPTER 17

Model Verification and Validation

17.1 Introduction

Model verification and validation are processes for assessing the quality of a model's forecasts. The terms verification and validation are often used interchangeably, but they actually represent two distinct parts of the process of model evaluation. *Verification* is concerned with assessing whether a model's algorithms and equations are complete and implemented properly. It also assesses whether there is sufficient and proper input data for the initial and boundary conditions, and the appropriateness of the data assimilation used. One of the main goals during verification is to eliminate any bugs or coding errors present in the modeling system and to test how robust it is. *Validation* is concerned with determining how well a model accurately represents a real atmosphere. Validation is an iterative process, comparing the model output to actual atmosphere behavior, and evaluating the discrepancies to improve the model. This process is repeated until model accuracy is judged to be acceptable. Accuracy is usually defined as the average degree of correspondence between individual forecasts and observations[1], although modifica-

---

[1]Murphy, A. H., 1988: Skill scores based on the mean square error and their relationships to the correlation coefficient, *Mon. Wea. Rev.*, **116**, 2417-2424.

tions of this definition will be used as needed throughout this chapter.

Unfortunately, the term verification is also used in the operational meteorology community as a measure of an operational weather forecasting modeling system's ability to predict specific predefined atmospheric states. In this usage it has the same meaning as what we refer to as validation above.

In many cases where model output is used to provide weather warnings, or will be used for making policy decisions, only verified and validated models are allowed to be used. A typical example would be air quality models that are used for implementing and assessing air quality standards for communities. Such models must be certified and approved by government agencies. The term *certified model* sometimes is used to describe an officially recognized acceptable model for a given purpose. Because of the lengthy process of verification and validation, it is often difficult for more advanced, recent models to be officially accepted for use.

Throughout this book we have discussed the reasons for model errors, which are summarized here:

- Representation of a continuous medium such as the atmosphere using a discrete grid.

- Numerical approximation of the governing equations.

- Incomplete initial and boundary conditions.

- Errors in initialization data.

- Parameterization of subgrid-scale processes.

There is also a limitation to the variables that can be produced as direct model output. For example, the occurrence of a thunderstorm is a very important variable, but it must be derived from the direct model output. Postprocessing techniques have been developed to add value to the direct model output in term of producing needed derived variables, and to reduce error in order to improve the accuracy of output for points within the grid. Some of these techniques will be described in following sections.

17.2   Verification

The main goal of model verification is to ensure that valid algorithms are being used within the model, and to 'de-bug' the code, ridding it of typographical and coding errors which are invariably present in any computer code of even modest length. During verification, each sub-module (the advection routine, dispersion routine, radiation parameterization, etc.) is first checked and tested separately. This is often accomplished by starting with canned, well-established test cases that have known input and output values.

After all modules have been verified separately, the modeling system is verified in its entirety, again using canned test cases. This part of the verification process is similar to validation, except that here we are mainly assessing the model's ability to yield 'reasonable' answers of the proper sign and order of magnitude using only a few test cases, and ensuring that there are no bugs causing the model to become unstable or to terminate prematurely. During the verification step the general accuracy of the model is not assessed.

17.3   Validation

After verification the model should be free of bugs and coding errors, and should yield 'reasonable' values for the predicted variables. The model is then validated to assess how well it performs. The main criteria that are assessed during the validation process are 1) *accuracy*, 2) *biases*, and 3) *reliability*. For models that are used for forecasting purposes, additional factors such as *skill* and *utility* are also assessed.

Accuracy is the difference between the forecast values of variables and their observed or analyzed values. Accuracy can be quantified in many different ways, but one common method of quantification is to average the squares of the differences of the variables over the model domain.

In many instances the model variables will be *biased* toward either over-prediction or under-prediction, and these biases will be established during the validation stage. The biases may be the result of the subgrid-scale parameterizations. The biases may not be consistent over the model domain, but instead may differ depending on the geographic location and season. Once the biases are established, efforts can then be made to reduce them by modifying appropriate submodules of the modeling system, or by applying postprocessing corrections to the model output (see Chapter 18).

Reliability is the term used to relate the correlation of the forecast values to the average values for specific observed variables.

Skill is a comparison of the model forecast with some reference forecast. Usually the reference forecast will be either persistence or climatology, but another modeling system can also be used as the reference. If the accuracy of the model is better than the accuracy obtained by the reference forecast, then the model is said to have skill. If the model performs worse than the reference forecast, then the model is said to have no skill.

Utility refers to how useful the model forecast is for decision making in a specific application or industry. For example, an electric utility company using wind farms may use a model to forecast wind conditions, which can be used to determine usage and load. In this case the utility of the model would be a measure of how helpful the model output is in making decisions regarding operation or maintenance of the wind farm.

Although the validation stage should mostly be an objective process, there may still be some subjectivity in the definitions of what constitutes skill or utility. For example, the reference standard chosen for comparison when determining model skill is chosen subjectively. The utility of the model is also subjectively dependent on the needs of the user of the forecasts. That is, the purpose of the forecast will have a significant influence on what is deemed a "good" or "bad" forecast. For example, a forecast that supports a solar power prediction system would focus on cloud forecasts and would not be influenced by poor wind forecasts, while a wind power prediction system would focus on wind speed and not be directly influenced by poor cloud forecasts.

Any change in the code of a model in effect makes it a new model that needs to be revalidated. The most efficient way to validate before releasing a new or updated model is to obtain and establish a quality controlled historical dataset that covers a long period, that is well studied and understood. A rigorous validation process would consist of comparisons of one, two and three dimensional variables, spatial patterns and time series over all seasons.

17.3.1 Methods of validation

The validation process should be as objective as possible and based on a statistical assessment. A standard statistical assessment consists of calculations of *mean error* (also called *bias*), *mean absolute error*, and *root mean square error*. In addition to standard statistics, forecast skill scores can also be used.

An important aspect of model validation is sample selection. There is no perfect sampling method, and so the error statics generated during validation should be considered as only approximations to the actual model error. One method to improve sample selection is to separate the sample into regimes, or subsets of samples, that have similar dynamical and physical properties. Examples are dividing the sample by season or synoptic regime (e.g., high zonal versus low zonal flow patterns).

The most basic form of statistical analysis of forecast accuracy is the use of an empirical joint distribution comparison made between a set of forecasts and the relevant observations.[2] There are several statistics that are useful for quantifying the error and accuracy of a model. The most basic of these is the mean error (ME). There are several alternate names given to this statistic by different authors, such as *mean bias error* or just *mean bias*. The bias or error of a forecast is the difference between the forecasted value and the observed value. In equation form the error at a model grid point is

$$\varepsilon = f - o \qquad (17.1)$$

where f is the model forecast value and o is the value of the observations interpolated to the grid point. If $\varepsilon = 0$ the forecast is unbiased at that particular grid point. The bias may very well be grid-point dependent.

The bias for a single forecast is not very useful in determining the typical accuracy of a given model at a grid point. It is more useful to calculate the average bias or mean error, $\bar{\varepsilon}$, for a large number (N) of forecasts,

$$\bar{\varepsilon} = \frac{1}{N} \sum_{n=1}^{N} \varepsilon_n. \qquad (17.2)$$

---

[2]It is also sometimes useful to compare the forecast of one model with the forecast from another separate model for which the error statistics are known. In essence this treats the reference model as the observations or truth.

The mean error still has limited utility since a mean error of zero does not imply that any individual forecast will have a low error. For example, if half of the forecasts had a large positive bias while the other half had a large negative bias, the mean error would be near zero.

In addition to mean error, other useful parameters are the *mean absolute error (MAE), mean-squared error,* and the *root-mean-squared error (RMSE)*, given by

$$MAE = \overline{|\varepsilon|} = \frac{1}{N} \sum_{n=1}^{N} |\varepsilon_n| \tag{17.3}$$

$$MSE = \overline{\varepsilon^2} = \frac{1}{N} \sum_{n=1}^{N} \varepsilon_n^2 \tag{17.4}$$

$$RMSE = \left(\overline{\varepsilon^2}\right)^{1/2} = \sqrt{\frac{1}{N} \sum_{n=1}^{N} \varepsilon_n^2} \tag{17.5}$$

(note that $RMSE = \sqrt{MSE}$). These are all more useful indicators of model accuracy than is mean error; however, they do not give any indication as to whether the model is over or under predicting. Values of MAE and $RMSE$ range from a perfect score of 0 all the way to infinity. MAE is a good way to quantify forecast error in a time series analysis. $RMSE$ is influenced more strongly by large errors and can be a useful measure of how well a model captures extreme events. The mean error, mean absolute error, and root-mean squared error have all been defined for a specific grid point. Their values may vary widely over the model domain. Domain-averaged values of these parameters can also be calculated and are sometimes useful.

In addition to the objective statistical measures of accuracy that involve point comparisons, subjective validation methods are also useful for evaluating the performance of a given model. Subjective methods involve visually estimating the accuracy of patterns for both continuous and discontinuous variables. Traditional validation scores do not always agree with the subjective assessment of the forecasts, which leads to the question of, "which is the correct way to assess forecast accuracy?"[3] There is no clearly correct

---

[3] Ahijevych, D., E. Gilleland, B. G. Brown, and E. E. Ebert, 2009: Application of

answer to this question. For continuous variables such as pressure, temperature and dewpoint, an evaluation of gradients and Laplacians (the gradients of gradients) are valuable parameters that can be objectively calculated using the standard statistical methods already mentioned. For discontinuous variables, such as precipitation and cloud coverage, other methods are needed.

Several spatial forecast validation methods have been developed that are suited for high-resolution precipitation forecasts.[4] These methods can be grouped into four broad categories: 1) neighborhood; 2) scale separation; 3) features-based, and 4) field deformation. They give accuracy credit to a forecast that shows skill but does not necessarily match an observation at any particular grid point and time. Each of the spatial forecast methods is able to detect bias in a model and account for the spatial coherence of precipitation. The features-based and field deformation methods are also able to diagnose displacement errors of precipitation features. The field deformation method is a good approach for capturing errors in aspect ratio (the proportional relationship between the longitudinal and latitudinal extent of a meteorological feature).

17.3.2 Contingency tables

Contingency tables provide method of displaying the frequency distribution of multiple variables. Binary, or 2×2, contingency tables are useful for assessing the accuracy of forecasts for parameters that have only two possible outcomes, such as rain versus no rain. Table 17.1 shows an example of a binary contingency table, representing the four possible outcomes between the observed and forecast parameters. The parameters in the table represent the number of instances for which: *a*) the parameter was forecasted to occur and did occur; *b*) the parameter was forecasted to occur and did not; *c*) the parameter was not forecasted to occur and did; and *d*) the parameter was not forecasted to occur and did not. Outcomes *a* and *d*

---

spatial verification methods to idealized and NWP-gridded precipitation forecasts, *Wea. Forecasting*, **24**, 1485-1497.

[4]Wernli, Heini, Christiane Hofmann, Matthias Zimmer, 2009: Spatial forecast verification methods intercomparison project: application of the SAL technique, *Wea. Forecasting*, **24**, 1472-1484.

represent correctly forecasted events and non-events,[5] while b and c represent incorrectly forecasted events and non-events.

Table 17.1: Binary contingency table.

	Observed	Not observed	
Forecasted	a	b	$a + b$
Not forecasted	c	d	$c + d$
	$a + c$	$b + d$	$N = a + b + c + d$

Several statistical quantities can be derived from the contingency table. The most basic quantity is *percent correct (PC)*, which represents the fraction of forecasts that were correct,

$$PC = (a + d)/N. \tag{17.6}$$

For extreme events, such as hail or tornado occurrence, PC is not very meaningful because there will be a large number of forecasts made for which the event was not predicted to occur, and that were correct (parameter d in the table). This will skew the PC value towards unity even if there are several missed forecasts that predicted no event, and for which the event occurred. This shortcoming is compensated for by three other validation parameters described as follows.

The *probability of detection* or POD is defined as the fraction of observed events that were forecast correctly, and is calculated as

$$POD = a/(a + c). \tag{17.7}$$

It only evaluates those forecasts that predicted the event to occur. The *false-alarm ratio* or FAR is the fraction of forecasts that predicted the event to occur and for which the event did not occur. It is given mathematically as

$$FAR = b/(a + b). \tag{17.8}$$

Neither probability of detection nor false alarm rate is useful in isolation. For example, if every forecast predicted the occurrence of

---

[5]A 'non-event' is simply an instance where the phenomenon in question did not occur. If no severe thunderstorms are forecasted, and none occur, that is a correctly forecasted non-event.

a tornado, then the probability of detection would be a perfect value of unity. Likewise, if no forecast ever predicted a tornado, then the false-alarm rate would have the ideal value of zero. A means to combine the information contained in the probability of detection and the false-alarm ratio is the *threat score (TS)*, calculated by

$$TS = a/(a + b + c). \tag{17.9}$$

The threat score excludes those forecasts that did not predict event occurrence and which were correct. A threat score of unity represents a perfect forecast system that predicted one-hundred percent of the positive events correctly and had no false alarms.

The bias can also be calculated from the data in the contingency table as

$$\varepsilon = (a + b)/(a + c). \tag{17.10}$$

If the forecast system is unbiased, then $\varepsilon = 1$. An ε less than unity represents a forecast system that underforecasts the event, while a value greater than unity represents an overforecast.

17.3.3 Skill score

Forecast skill is a measure of the relative accuracy of a set of forecasts with respect to some set of standard control, or reference. Possible references are climatology and persistence forecasts. The *skill score* is expressed as a percentage improvement over the reference forecast.

The basic skill score is computed by comparing either the mean absolute error or the mean-squared error of the forecast values versus the reference values. Denoting the reference values as R, the forecast values as f, and the observed values or 'truth' as T, the mean-squared errors of the reference and the forecast are

$$MSE_R = \frac{1}{N} \sum_{n=1}^{N} (R_n - T)^2 \tag{17.11}$$

$$MSE_f = \frac{1}{N} \sum_{n=1}^{N} (f_n - T)^2. \tag{17.12}$$

The skill score is then calculated as

$$SS = 1 - MSE_f / MSE_R. \tag{17.13}$$

A perfect forecast would have no mean-squared error and therefore a skill score of unity. A forecast that has greater mean-squared error than the reference would have a skill score that is negative. There are many variations of the skill score method that have been developed for specific applications.

17.3.4   Reliability diagrams

Many forecasts are probabilistic in nature, such as precipitation forecasts. An approach to evaluating probabilistic forecasts is the *reliability diagram*, in which the forecast probability is plotted on the *x*-axis and the verifying frequency of occurrence of the values of probability is plotted on the *y*-axis. The following is an example following Wilkes.[6] If a model probability forecast were validated for a given variable or event such that all of the 0, 20, 40, 60, 80 and 100 percent forecasts were validated against the percent occurrence of the variable, a reliability diagram of observed relative frequency as a function of forecast probabilities could be constructed with the observed relative frequency as the *y*-axis and the model forecast probability as the *x*-axis. Given the following data,

Forecast probability	0	20	40	60	80	100
Observed frequency	5	15	25	65	90	99

the corresponding reliability diagram appears in Fig. 17.1. Ideally the forecast probability and observed frequency curves would be identical. The plotted reliability diagram shows a tendency of the model to overforecast in the 20 to 40 percent probability range and underforecast in the 60 - 80 percent probability range.

17.3.5   Kinetic-energy spectra

As models run at higher and higher resolution, there is an increasing need to find alternative ways of validating models. One proposed method is to compare the model's kinetic-energy spectra with observed kinetic-energy spectra as a validation of the ability of a mesoscale model to capture high resolution phenomena. In stud-

---

[6]Wilkes, D. S. 1995: *Statistical Methods in the Atmospheric Sciences*, Academic Press, 648 pp.

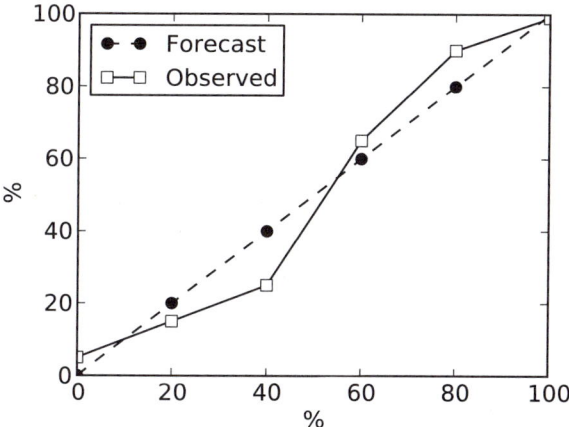

Fig. 17.1: Reliability diagram plotted for data in the table shows a tendency of the model to over forecast in the 20 to 40 percent probability range, and under forecast in the 60 to 80 percent probability range.

ies[7] it was found that the kinetic energy spectra computed using the Weather Research and Forecast (WRF) mesoscale model at near-cloud-scale matched the observational spectra well, including the periods of transition from turbulent to steady state flow. In this method the model spectral decay was compared with the observed wave spectra at the highest resolved wave numbers in the model. The departure of the model wave spectra from the observed wave spectra is used to define the model's effective resolution. This technique offers the possibility of determining 3D model accuracy in a more complete way than traditional accuracy methods.

17.3.6 Interpretation of validation statistics

The mathematics behind model performance statistics is generally straightforward. However, performance statistics themselves can be difficult to interpret. Errors that are the result of nonsystematic (random) events can sometimes mimic systematic errors, making forecasts look artificially better or worse than the true error or accuracy. Figure 17.2 provides an example of increasing nonsystematic error of a temperature anomaly forecast with increasing forecast

---

[7]Skamarock, W.C. 2004: Evaluating mesoscale NWP models using kinetic energy spectra, *Mon. Wea. Rev.* **132**, 319-332.

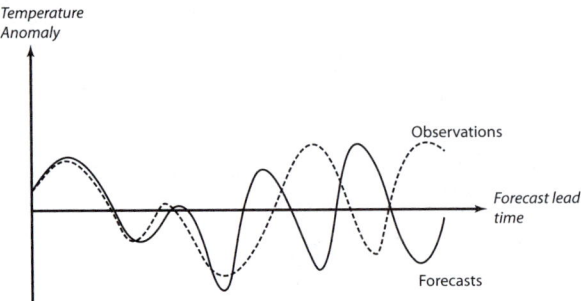

Fig. 17.2: A schematic picture of a medium range forecast (solid black lines) and the verifying analysis (dashed black lines). The forecast anomalies have about the same magnitudes as the verifying anomalies, but they are out of phase. This will yield a tendency for positive anomalies to verify against less positive or even negative anomalies, and negative anomalies to verify against less negative or even positive anomalies.

lead time. Early in the forecast period the error could be mistaken as a systematic error of an otherwise "good" forecast.

Note that in Fig. 17.2 the model-forecast temperature anomalies have essentially the same intensity and frequencies as observed temperature anomalies, but as the lead time increases, the artificially forced anomalies tend to be displaced in time. A model that has less forecasting skill as the lead time of the forecast increases shows up as a phase error in the anomaly diagram.

One difficulty of properly interpreting forecast accuracy using statistics is that different evaluation methods can give very different impressions of forecast quality. Paradoxically, it is not unusual for some measures of forecast error to be increased for a modeling system after upgrades in model physics or parameterizations are made. The reason is that while the improved model will often give more realistic simulations of the structure of small features, the phase and spatial error may be larger.

## CHAPTER 18

Postprocessing Enhancement of Model Data

18.1 Introduction

Model output can be used directly to predict atmospheric variables. However, there are limitations to the direct use of model output. One key limitation is that not all parameters of interest for a particular model user are produced by the model, and so must be derived from the direct model output during postprocessing. An example of such a parameter is radar reflectivity. Another key limitation is accuracy; all models have biases and errors. Postprocessing techniques have been developed to remove the systematic biases and reduce error. In practice, only systematic error can be reduced statistically from the model output.

18.2 Statistical Methods

Statistical methods have been created in an attempt to reduce error and improve model accuracy based upon the correlations between the three-dimensional fields produced by the models and a set of verification observations and/or the climatological conditions for specific locations. Two basic statistical approaches are used:

perfect-prog[1] and model-output statistics (MOS).

Both approaches are analogous to what an experienced human forecaster does when using model output as an aid to making forecasts. In the case of the human forecaster, comparisons are made between the model forecasts and a set of verification observations. The forecaster will then apply a correction to the model forecast based upon a subjective evaluation of biases and error in the past model forecasts. Experienced forecasters will even make note of different biases and errors based upon differing synoptic scale patterns or regimes, and apply subjective regime based corrections. In this way an experienced forecaster can improve upon the raw model forecast.

In the perfect-prog and MOS methods, multiple regression equations are calculated from a large sample of predictor variables. There are key differences between the two methods. The most important difference is the fact that the perfect-prog method develops the regression equations using only observed data, while the MOS method uses both observed and model data. Thus, the MOS method can be used to remove biases and systematic error from the direct model forecasts, while the perfect-prog method cannot.

In addition to improving the accuracy and usefulness of model output, the statistical methods can also be used to derive forecasts for basic weather elements, such as visibility and lightning, that are not directly calculated in the model or directly derivable from the direct model forecast variables.

18.2.1   Perfect-prog method

The perfect-prog technique uses past observations and climate data to develop a set of regression equations relating two or more meteorological variables. The regression equation would have the form of

$$\Phi = f(\phi_1, \phi_2, \phi_3, \cdots, \phi_n), \qquad (18.1)$$

where the ϕ_n are the model output variables that are used to diagnose the parameter Φ. The regression equations are formulated from a purely statistical standpoint, though the *predictors* (the ϕ_n) are expected to have some physical relationship to the *predictand* (Φ). Once an equation of the form (18.1) has been established, it can then be applied to model output, using the model-predicted values

---

[1]Short for perfect prognosis.

of the ϕ_n to predict the value of Φ. The name 'perfect-prog' comes from the assumption that the model output for the ϕ_n is completely accurate. Model biases are not factored into the prediction of Φ. The regression equation (18.1) can be used with the output from any numerical model whose output includes the predictors.

18.2.2   Model-output statistics (MOS)

The technique of model-output statistics, which is usually just written as *MOS* and pronounced 'moss', is similar in derivation to the perfect-prog technique, with one major exception. Unlike the perfect-prog technique, which uses observed and climatological values of the predictors to formulate the regression equation for the predictand, MOS uses the output of the model itself for the predictor values. The advantage of this method is that the MOS regression equations account for model biases.

MOS is very popular, and used extensively for adjusting operational model output for point forecasts. In order for MOS guidance to be robust and reliable, a long period of record for both model forecasts and observations is necessary in order to develop the regression equations. Ideally, the length of the period of data record should be several years of historical records, which can be difficult to obtain. A further complication is that the MOS regression equations are tied to a specific version of a specific model. Anytime the model equations or parameterizations are changed or updated a new set of MOS regression equations must be developed.

MOS objectively does what an experienced forecaster does when adjusting a model forecast based upon perceived model biases from past forecasts. For example, a forecaster may subjectively adjust a temperature forecast downward based upon the fact the model forecasts had been high in the past. The problem is that the amount of adjustment made to the model forecast is subjective because it is not based upon an established relationship. However, MOS will objectively establish a statistical relationship that will consistently account for the model bias such that it will be consistent in making adjustments to all temperature forecasts.

There are several approaches to using MOS for improving numerical forecasts. The standard approach is to use a long period of record as a training set with one set of calculated regression equations using a single model for the entire period. For most cases the basic MOS system produces more accurate forecasts than the raw numerical forecasts. But there are times during transition months when this approach can make the final forecast worse than the raw numerical forecast. Another problem when trying to provide forecasting services to users in remote locations is the lack of a long period of record for use as a training period.

To address the long-period-of-record problem, an approach that is typically referred to as *dynamic MOS* uses only a recent history, typically a few weeks, as the training set along with a continual recalculation of the MOS regression equations. This approach solves the problem of using MOS for stations without a long period of record, and it tends to help reduce the problem of MOS making the final forecasts less accurate than the raw numerical forecast. However, on average the dynamic MOS approach does not reduce the model bias as much as the standard MOS.

A third approach is the *regime MOS*, or *stratified MOS* method, which uses a subset of cases from a specific weather regime (e.g., high zonal, low zonal, blocking, winter, etc.) to build the training set during a defined period. If the regimes are selected correctly, this approach can very effectively reduce bias for those situations where standard MOS performs poorly. However, the need for a long period of record becomes even more of a problem because a reasonable number of cases are needed to define the regimes and create the training sets from the entire period of record.

To find a solution that addresses both the period of record problem and the tendency of standard MOS to make extreme forecasts worse, a hybrid *dynamic-regime MOS (DRMOS)* has also been adapted for use in operational forecasting. This method uses the regime approach to define a subset of training days and continually recalculates the regression equations; however, the regimes are dynamically redefined each day based on how well the current regime matches an archived sample. The matches are based upon a set of pre-defined "matching parameters," but there are no pre-defined sets of regimes. A new training sample is created each day based upon the sample of historical cases that most closely matches the forecast days' matching parameters.

18.2.3   Comparison of perfect-prog and MOS

Though both perfect-prog and MOS are similarly derived, they will not give identical results. Each has strengths and weaknesses. The advantages of the perfect-prog approach are:

- There is no need to have a historical record of model forecasts in order to derive the regression equation, since only observations are used as the predictors.

- If model forecasts improve, so will the perfect-prog forecast.

- Correlations between predictors and predictands tend to be very high.

Disadvantages to the perfect-prog approach are:

- Does not account for model biases.

- Cannot use model-derived parameters such as vertical velocity or cloud liquid-water content as predictors.

In contrast, the advantages to the MOS approach are:

- Can account for systematic model biases and errors.

- Can use model-derived parameters such as vertical velocity or cloud liquid-water content as predictors.

- Shows better skill than perfect-prog for longer-range forecasts.

Disadvantages of MOS are:

- Requires a long period of historical model data in order to develop the regression equation.

- The regression equation is model-dependent, meaning any changes in the model require new regression equations to be developed.

- The relationships between the predictors and the predictand weaken with time as the model error increases.

18.2.4   Artificial neural networks

Artificial Neural Networks (ANNs) take a somewhat different approach for improving numerical forecasts. They are similar to MOS techniques in that they produce an adjustment factor that is designed to eliminate bias and reduce model forecast error for a point. The difference is in how the adjustment factor is calculated. In a broad sense this is just a different way to approach calculating the MOS regression equations.

The standard MOS approach is one of analyzing the model forecast and observed conditions to find correlations from a historical data set, and then calculating an adjustment factor and applying the adjustment to future forecasts. The ANN approach is an algorithm that is trainable, allowing the program to learn which variable relationships are important and using this learned knowledge to adjust the regression equations to reduce model bias through an iterative process. The model adjustment equations are continually refined as the ANN gains knowledge about the relationships through assigning weights to the adjustment equations. Keep in mind that the ANN approach can be combined with other techniques such as MOS in order to achieve a greater reduction in error of the model forecast than using any single technique alone.

18.3   Ensemble Techniques

The ensemble forecasting method is an attempt to systematically evaluate the uncertainty in numerical forecasts due to: 1) the errors caused by imperfect initial conditions and sensitivity to the initial conditions; and 2) the errors caused by the approximations made in the model formulation. There are two main approaches to the ensemble method.

The first approach is to perform multiple runs of the same numerical model, only with each run having slightly perturbed initial conditions. If the output from the models is closely grouped around some mean value, then there is confidence that the actual future state of the atmosphere will lie somewhere within the cluster of ensemble member output. If the output is not closely grouped, then it gives a sense of the uncertainty of the model forecasts.

In the second approach an ensemble is created from single runs of completely separate models, some or all of which may be from different agencies or organizations. If there is a tight grouping representing model consensus, then confidence can be had that the future state of the atmosphere will lie somewhere within the cluster of ensemble member output.

The output from the ensemble members is usually displayed in either a *spaghetti plot* or a *postage-stamp* plot. In a spaghetti plot a certain isobar, isohypse, position of the jet stream, or other representation of the height field from each ensemble member is overlaid on a map. In the postage-stamp plot a thumbnail representation of a map from the ensemble member is displayed on a single page or panel. Quantitative statistics can also be computed, such as the standard deviation of a particular variable among the ensemble members.

There is no one best approach to perturbing or selecting model inputs and options to configure an ensemble. A major reason for this is that there is no completely objective way to calculate the amount to perturb the inputs to the members of the ensemble. Another significant reason is that the "best" perturbation to use in an ensemble is application dependent. For example, an ensemble that is configured to produce an optimum spread of forecasts for precipitation would not typically be the configuration that would provide the optimum spread of forecast for atmospheric transport and dispersion. The best perturbations for precipitation forecasts must cause a large spread in forecasts of clouds and precipitation among the ensemble members. However, for transport and dispersion forecasting there needs to be a large spread in predictions for low-level wind direction and atmospheric boundary layer depth.

## CHAPTER 19

Special Applications

19.1   Introduction

In this final chapter we survey some of the specialized applications of atmospheric numerical modeling. There are many models that are configured to optimize forecasting for special meteorological and ocean related phenomena, such as tropical cyclones, waves, storm surges, forest fires, agriculture, and renewable energy.

In many cases the specialized model takes information from larger-scale models, such as wind and temperature, for initialization, but no information from the specialized models feeds back to the larger-scale model. In some case the models are dynamically coupled with the larger-scale models, so the results from the specialized model are fed back to the parent model. For example, some ocean models are coupled with an atmospheric model.

19.2   Aviation Weather

Accurate weather forecasts are essential to aviation, for both safety and economic reasons. Some of the weather elements needed by the aviation community, such as temperature and dew point, are direct model outputs, but many of the most significant weather parameters that impact aviation are not directly available from atmospheric numerical model. Variables such as ceiling heights, visibility, thunderstorms (lightning), hail, turbulence, up and downdrafts, low-level wind shear and icing, can cause severe damage to an aircraft during take-off or landing and in flight. Specialized models that use inputs from larger-scale models are often used to provide forecasts for these variables. Volcanic ash can be a significant problem for aviation, as an aircraft can lose engine power within ash clouds, so there are models tailored to predict ash movement. Also, on a day-to-day basis, airliners use the output of air traffic route models that are designed to take advantage of the inflight winds in order to predict and improve fuel economy.

19.3   Wildland Fire Models

Weather conditions such as wind, precipitation and humidity are key to the development and control of wildland fires. Thus, the forecasting of these elements is essential for preventing as well as controlling wildfires. Models have also been developed to predict the direction and speed of wildland fires. More sophisticated wildland fire models have a two-way influence, whereby the fire itself can impact the convection, condensation, and wind field in the atmospheric model.

19.4   Maritime Models

There are many types of specialized ocean and ocean-related models that are used to describe and predict maritime phenomena such as ocean currents, ocean temperatures, ocean waves, tropical cyclones and storm surges. Models are used to depict the sea state and predict the evolution of wind-driven waves. These wind-wave models use the wind forecasts from atmospheric models as input. In a similar fashion, storm surge models utilize wind and pressure forecasts from atmospheric models as input, and also take into account coastline shape and bathymetry to estimate surge height and inland penetration.

19.5 Power Forecasting

Electrical and gas utilities rely on weather forecasts to anticipate customer demand, which can be strongly affected by temperature. Another use of specialized utility forecasts is to help predict the generation of power from intermittent renewable sources of power, such as wind and solar. By understanding power load and generation, utility companies can purchase additional supplies of power or natural gas before the price increases, or in some circumstances, in order to prevent brownouts and blackouts.

19.5.1 Renewable power forecasting

In order to achieve optimal wind and solar power generation and use, two types of meteorological information are required. First, there is a need for high-resolution spatial maps of the wind and solar irradiance climatology. The maps are used for optimally siting wind and solar power generation facilities. Second, there is a need for short-term (0–48 hr) forecasts of the hourly wind power and solar power output. The forecasts are used to 1) improve scheduling and dispatching of generation and transmission resources and 2) optimize wind and solar energy buying and selling strategies in deregulated markets, thereby increasing the reliability of renewable energy as a power source.

There are significant obstacles to reliable wind and solar mapping in most areas of the world because of the lack of availability and accuracy of observations of wind speeds and solar irradiance. The solution to this problem has been the development and refinement of specially configured models in combination with satellite observations and sophisticated stochastic methods that produce highly accurate wind and solar climatology maps with a resolution as high as one kilometer.

In addition to mapping, there is a need for very accurate and high-resolution wind and solar insolation[1] forecasts. For wind power, the wind speed at the height of the turbine, which is usually about 50 meters above the surface, is needed. For solar-power, high-resolution cloud-cover forecasts are required. To address these needs, highly sophisticated dynamical-statistical modeling systems

---

[1]Insolation is the power flux of incoming solar radiation, and its name is in fact derived from IN-coming SOL-ar radi-A-TION.

are used. These combine refined statistical MOS[2] output in order to produce accurate wind speed and insolation forecasts. In a similar fashion, solar forecasting procedures are based upon the use of an ensemble of model simulations adjusted by various MOS techniques.

19.5.2   Power ramps

One of the major challenges for both wind and solar power forecasting is to forecast large changes in power output over very short periods of time. These changes in power are referred to as *power ramps*. Down ramps are very important, because if unpredicted there could be a serious shortage of available power resulting in higher production costs, brown outs, and even black outs of power. Up ramps are also important because overage of power on the system can also cause serious system overloads. Both types of power ramps need to be forecasted accurately. Numerical models combined with probability of occurrence methods are used to help grid managers plan for ramp events.

19.5.3   Hydropower forecasting

Real-time operational hydrological forecasts are needed for hydroelectric power generation, water resource planning, and to provide warnings of possible water-related disasters. Although much more reliable than wind and solar power, hydropower also has variability that is dependant on weather, primarily precipitation and snow melt.

There are several modeling approaches to real-time hydrological forecasts, each having strengths and weaknesses. Most methods used atmospheric model forecasts of temperature and precipitation as input to a hydrological model. The *distributed model* is a physics-based, spatially-distributed hydrology model that explicitly represents hydrological processes, and includes the effect of diverse topography and heterogeneous subsurface conditions on the downslope redistribution of subsurface moisture. Typically, precipitation observations from surface rain gauge data and radar-derived precipitation are used to provide input into the distributed model, before the beginning of the forecast period. Also, a mesoscale atmo-

---

[2]Model-output statistics. See Chapter 18

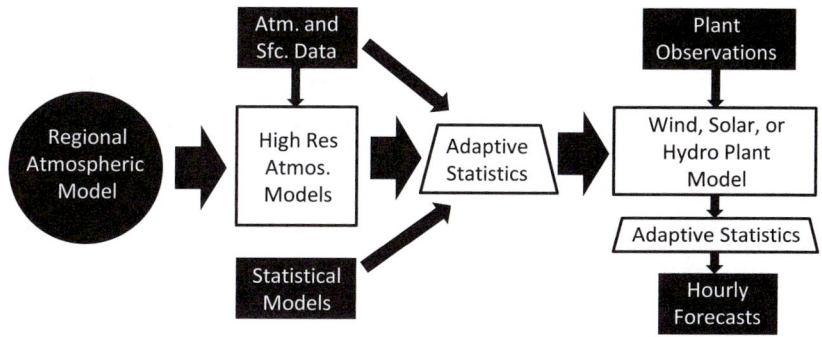

Fig. 19.1: A schematic of a model-based renewable power forecasting system.

spheric model provides bias-corrected precipitation and temperature forecasts which drive the distributed model during the forecast period. The distributed model output can be aided by independent statistical models and MOS adjustments to model forecasts.

19.5.4 Components of renewable power forecasting modeling systems

The basic renewable power modeling system is shown in Figure 19.1. The regional weather model provides the initial conditions for the higher-resolution model domain, as well as updated boundary conditions. The output from the higher-resolution models is combined with additional meteorological parameters, as well as the output from statistical models, and then processed using adaptive statistical methods.[3] The results are then used as input into the energy-specific (wind, solar, hydro) model, which also ingests additional data from localized observations. The output from the energy-specific model is again processed using adaptive statistics, to produce the final power forecast.

---

[3]Adaptive statistics refers to a type of statistical analysis whereby the statistical methods used can change during the course of the analysis, based upon preliminary results.

19.6   Agricultural Forecasting

To reduce costs and increase productivity, the agricultural industry relies heavily on weather forecasts for variables such as temperature, precipitation and humidity to decide what work or action to take on any particular day. For example, growers need to plan in advance to react effectively to frosts and freeze conditions, periods of prolonged dry or wet weather, etc.

Specialized numerical models have been designed to accurately represent the conditions at the canopy level for various crops, based upon the leaf radiative and transpiration characteristics. In addition to using direct model-produced variables, antecedent observations and model output is used as input to various agricultural models that predict parameters such as the timing of the development stage of harmful insects and crop diseases, so that pesticides and herbicides may be applied at the proper time. These agricultural models not only decrease costs to the grower, but also decrease application of unneeded chemicals.

19.7   Historical and Climate Data

Climatological data is very useful in many applications. However, there are many locations on the globe for which there are not adequate historical records for generating a climatology. In these instances it is possible to generate a synthetic climatology using *retrospective modeling*. Special models have been designed or adapted for this purpose.

Generating synthetic climatologies using numerical models actually has certain advantages, the first and most obvious being that it may be the only means of quickly generating any climate information for a particular location. Climatological statistics for many different parameters can be generated, including parameters that are not part of a standard, archived weather observation. However, due to a lack of verification data it is not always possible to quantify the representativeness of climatologies produced by the retrospective-modeling technique. Another advantage of synthetic climatologies is that they explicitly take into account the effects of surface characteristics such as terrain, land-water distribution and surface roughness and their influence on the local scale environment.

When using a numerical-modeling approach to simulate climate data it is important to select a period of record for which initial and lateral boundary conditions are available. If observed data are available, it is important to assimilate that information as well to help reduce any model drift. Care must be taken during the time of data simulation if time series information is to be used from the data produced by the climate simulations. If data assimilation causes imbalances, there may be discontinuities in the time series. For example if the rawindsonde has a dry bias, and if this data is assimilated without accounting for this dry bias, then clouds, precipitation and temperature will all be greatly impacted during the data assimilation and spin up times.

19.8   Sensor Design, Performance, and Placement

Numerical models are often used to help design and determine the performance characteristics of new and existing sensors. Models can also help determine whether a specific sensor, or placement of a sensor, will help improve a model forecast. There is no point in designing or deploying a sensor if the information derived from it does not add value to a forecast or model simulation.

19.8.1   Sensor design and performance

There are two types of experiments involving the use of numerical-modeling techniques for determining sensor design and performance. The first type of experiment is called Observing System Experiments (OSE), which is used for an observing system that is already in operational use. The OSE methodology consists of the following steps:

1. A control run is made which uses all of the available observational data for a typical operational forecast model run;

2. A model forecast experimental run is made that excludes the observation from a specific sensor type or location in order to evaluate the forecast performance as compared to the control forecast run;

3. A comparison of forecast skill is then made between the control and experimental forecast model runs.

In essence, OSEs are data-denial experiments. They reveal specifically what happens when an observation or set of observations is removed and not assimilated into a forecasting system. In this way the impacts of those observations of the forecast can be quantified.

The Observing System Simulation Experiments (OSSEs) is a type of simulation experiment that is closely related to the OSE. The OSSE is typically designed to investigate the impacts of proposed observing systems and their deployment strategies. OSSEs can also be used to investigate the effectiveness of data assimilation systems to handle new observational platforms by testing the impact of the potential observations on them. The information obtained from OSSEs is generally difficult, or in some circumstances impossible, to obtain in any other way.

The structure of an OSSE is very similar to an OSE. The main difference is that OSSEs are used as assessment methods for proposed observing systems and not existing systems. The methodology of an OSSE consists of the following steps:

1. Create a set of reference atmospheric states for the entire OSSE period using a high resolution, well formulated atmospheric model run without data assimilation. The set of reference state model simulations are typically called the Nature Runs (NR). The NR provide the reference standard from which the simulated observations are extracted and against which the OSSE assimilation experiments are verified;

2. Simulated point observations for the proposed observing system then need to be created using the NR. The observations extracted from the NR must include realistic error characteristics of the proposed observing system;

3. A control model run is made for which all available data from the current operational systems are assimilated. To be effective as a control standard, the control run should be done with an NWP model that is different from the NR model;

4. After the control run is made, a series of experimental runs are made with the control run model that assimilates various numbers and locations of the proposed observing system;

5. A comparison of forecast accuracy and skill is then made between the control and experimental runs.

19.8.2   Optimal sensor placement

Providing proper initial conditions to a numerical model is extremely important in order to produce accurate forecasts. However, it is extremely costly to deploy sensors. It is advantageous to know if placing an instrument in one location will make more of a positive impact on a model forecast than another location. The procedure for doing this has become known as *observation targeting*, meaning that the observation sites are selected to have the greatest positive impact on forecast performance for the locations of interest. Numerical-modeling methodologies have been developed to try to automate the process of selecting good locations for placing instruments. The procedure for selecting targeted observations is complex and somewhat subjective.

Prior to deployment, a model sensitivity study is conducted to determine the best location and if applicable the best type of observation platforms for deployment. The observation platforms are then deployed with forecasts made using two forecasting systems. One system makes a forecast that uses the targeted observations, while another system makes a forecast without using the targeted observations. The two sets of forecasts are then evaluated to determine the impact of including the targeted observations.

19.9   Computational Fluid Dynamics

Atmospheric numerical models are essentially a sub-branch of the field of computational fluid dynamics (CFD). The same basic equations that govern atmospheric dynamics and thermodynamics also describe flows in other fluids besides the atmosphere. Physicists and engineers use such models for simulating fluid flows at very high resolution. However, unlike the models we have described so far, they do not include parameterization schemes for processes such as radiation, cloud microphysics, cumulus convection and soil moisture. This limits their ability to simulate certain phenomena, such as thermally driven circulations. However, limiting the parameterization of key processes reduces the computational requirements, thus allowing CFD models to run at much higher resolution than typical atmospheric models while using the same computational resources.

Very often it is a larger-scale atmospheric model that provides the initial fields that a CFD model needs to perform a simulation, in an analogous way that a global model provides initial conditions for a regional model. CFD models are effective in simulating wind flow and turbulence of small-scale blocking features such as hills, buildings and wind turbines.

A subclass of computational fluid dynamics models are *large eddy simulation (LES)* models. These models are designed to explicitly resolve features in a fluid flow above a certain size. Features smaller than the grid resolution must be parameterized.

Large-eddy simulations are often employed in the atmospheric sciences for a wide range of fluid flow regimes. They are used extensively for the study of boundary-layer turbulence and cloud-mixing processes. In these studies the larger turbulent eddies are explicitly resolved in the model, while the effects of the smaller eddies are parameterized or even scaled away as negligible. They are are also employed for studying planetary-scale features such as the Asian Monsoon, or the Madden-Julian Oscillation.

19.10   Commercial Sector Contributions to Modeling

We conclude with a brief discussion of the role and contributions of the commercial sector in atmospheric modeling. In the early history of numerical modeling the majority of the contributions were made by government agencies and research laboratories, and by government-funded research at academic institutions. The nascent development of a numerical modeling capability was by necessity complex, involved, and expensive, and it was natural that government agencies took a lead role in these projects. The goals of the first modeling efforts were the improvement of forecasts and warnings for commerce, agriculture, and also for military purposes.

However, modeling centers run by governments, either for civilian or military purposes, often do not meet the needs of industries with specific requirements. They lack flexibility, cannot adapt to new needs quickly, and are unable to tailor niche products for newly emerging industries. For such industries the private sector can tailor the model output to drive applications or models that meet the needs of a given user. Many of the specialized modeling activities that have been discussed in this chapter—one example being renewable power forecasting—have been developed or significantly advanced by commercial forecasting firms.

Answers and Hints for Selected Exercises

Chapter 2

Ex. 2.2: Hints: The natural log of the Exner function is

$$\ln \pi = \frac{R_d}{c_p} \ln p - \frac{R_d}{c_p} \ln P_R. \tag{A.1}$$

Differentiating (A.1) yields

$$\frac{1}{\pi} \frac{\partial \pi}{\partial x} = \frac{R_d}{c_p p} \frac{\partial p}{\partial x}.$$

Ex. 2.3: Hint: $(\partial p / \partial x)_p = 0$. Also, use the hydrostatic equation.

Ex. 2.4: Values of σ at the dots:

hPa	Left	Center	Right
200	0.00	0.00	0.00
250	0.06	0.11	0.08
300	0.12	0.22	0.17
400	0.25	0.44	0.33
500	0.38	0.67	0.50
650	0.56	1.00	0.75
800	0.75	——	1.00
1000	1.00	——	——

Ex. 2.5: Hints: Begin with

$$\left(\frac{\partial p}{\partial x}\right)_z = \left(\frac{\partial p}{\partial x}\right)_\sigma - \left(\frac{\partial z}{\partial x}\right)_\sigma \left(\frac{\partial p}{\partial z}\right)_x.$$

Use the hydrostatic equation, and also

$$p = p_0 + \tilde{p}$$

and

$$p_0 = (1 - \sigma)p_T + \sigma p_s.$$

An intermediate result is

$$\left(\frac{\partial p}{\partial x}\right)_z = \sigma \frac{\partial p_s}{\partial x} - \left(\frac{\partial \tilde{p}}{\partial x}\right)_\sigma + \rho g \left(\frac{\partial z}{\partial x}\right)_\sigma.$$

Also, it can be shown that

$$\frac{\partial p_s}{\partial x} = -\rho \frac{\partial \Phi_S}{\partial x}.$$

Chapter 3

Ex. 3.1: Values of dF/dx using $a = 0.4$, $b = -1.7$, $c = 2.3$, and $d = -5.0$ for $\Delta x = 1.0$:

x	Forward	Backward	Centered	Analytic
-87	9275.2	9487.4	9381.3	9380.9
33	1235.2	1159.4	1197.3	1196.9
50	2891.0	2774.4	2832.7	2832.3

Ex. 3.2: Values of dF/dx using $a = 2.6$, $b = -3.2$, and $c = 2.3$ for various values of Δx:

x	$\Delta x = 50$	$\Delta x = 10$	$\Delta x = 1$	Analytic
-87	-455.6	-455.6	-455.6	-455.6
33	168.4	168.4	168.4	168.4
50	568.7	568.7	568.7	568.7

The calculated values are identical to the analytical values because the error terms in the centered-difference scheme involve only third and higher order derivatives. For a quadratic function these derivatives are all zero. Therefore, regardless of the value chosen for δx, the error is zero.

Ex. 3.3: The first derivative at $x = 3$ with $\Delta x = 1$ is:

Forward	Backward	Centered	Fourth-order	Analytic
-49.745	-36.964	-43.354	-43.632	-43.634

The second derivative at $x = 3$ with $\Delta x = 1$ is:

Second-order	Fourth-order	Analytic
-12.7812	-12.8221	-12.8223

Chapter 4

Ex. 4.2: An intermediate result is

$$\lambda = \frac{1}{1 + \iota \sigma \sin k\Delta x}.$$

Multiply top and bottom by the conjugate of the denominator, and find the magnitude of the result.

Ex. 4.4: The expression for λ has the form

$$\lambda = -a \pm \sqrt{a^2 + 1},$$

where a is real-valued. Since $\sqrt{a^2 + 1} \geq 1$ for any real-valued a, the negative root will always have a magnitude greater than one.

Chapter 8

Ex 8.3: The matrix equation is

$$\begin{pmatrix} i\tilde{\omega} & f & -ikg \\ f & -i\tilde{\omega} & ilg \\ kH & lH & -\tilde{\omega} \end{pmatrix} \begin{pmatrix} A \\ B \\ C \end{pmatrix} = \begin{pmatrix} 0 \\ 0 \\ 0 \end{pmatrix}$$

where $\tilde{\omega} = \omega - \bar{u}k - \bar{v}l$.

Ex 8.4: For Part b., we need to find values of α and β such that the two equations

$$\frac{\partial(u + \beta\eta)}{\partial t} + \bar{u}\frac{\partial(u + \beta\eta)}{\partial x} + \alpha\frac{\partial(g\eta/\alpha + \beta Hu/\alpha)}{\partial x} = 0$$

$$\frac{\partial(u + \beta\eta)}{\partial t} + (\bar{u} + \alpha)\frac{\partial(u + \beta\eta)}{\partial x} = 0$$

are equivalent. Inspection shows that this will be true if

$$g\eta/\alpha + \beta Hu/\alpha = u + \beta\eta. \tag{A.2}$$

Matching up the coefficients on the u and η terms in (A.2) results in the two equations,

$$g/\alpha = \beta \tag{A.3}$$
$$\beta H = 1. \tag{A.4}$$

Solving (A.3) and (A.4) for α and β results in

$$\alpha = \pm\sqrt{gH}$$
$$\beta = \pm\sqrt{g/H}.$$

Ex 8.5: Equation (8.88) can be rewritten as

$$\lambda^2 \pm \iota\sigma\lambda - 1 = 0.$$

Using the quadratic formula this yields

$$\lambda = \mp\frac{\iota\sigma}{2} \pm \frac{\sqrt{4-\sigma^2}}{2}.$$

If $\sigma^2 \leq 4$ then $|\lambda| = 1$. If $\sigma^2 > 4$, then

$$|\lambda| = \left| \mp\frac{\sigma}{2} \pm \frac{\sqrt{\sigma^2 - 4}}{2} \right|,$$

and at least one combination of the signs will be greater than one.

For ease of notation and space considerations, we define the following:

$$
\begin{aligned}
[K_+^+] &= 1 + e^{\iota kd}, \quad [K_+^-] = 1 + e^{-\iota kd}, \\
[K_-^+] &= 1 - e^{\iota kd}, \quad [K_-^-] = 1 - e^{-\iota kd}, \\
[L_+^+] &= 1 + e^{\iota ld}, \quad [L_+^-] = 1 + e^{-\iota ld}, \\
[L_-^+] &= 1 - e^{\iota ld}, \quad [L_-^-] = 1 - e^{-\iota ld}.
\end{aligned}
\tag{A.5}
$$

Ex 8.6: Use solutions of the form

$$(u_i^n, v_i^n, \eta_i^n) = (A, B, C)\lambda^n e^{\iota kid}.$$

The coefficient matrix for A, B, and C is

$$
\begin{pmatrix}
(\lambda^2 - 1) & -2f\Delta t\lambda & -\frac{2g\Delta t}{d}[K_-^+]\lambda \\
2f\Delta t\lambda & (\lambda^2 - 1) & 0 \\
\frac{2H\Delta t}{d}[K_-^-]\lambda & 0 & (\lambda^2 - 1)
\end{pmatrix}.
$$

See (A.5). Setting the determinant equal to zero, the resulting equation can be put in the form of (8.88), where

$$\sigma^2 = \frac{8c^2\Delta t^2}{d^2}(1 - \cos kd) + 4f^2\Delta t^2.$$

Ex 8.7: The coefficient matrix for A, B, and C is

$$
\begin{pmatrix}
\iota\omega & f & -\frac{g}{d}\iota\sin kd \\
f & -\iota\omega & \frac{g}{d}\iota\sin ld \\
\frac{H}{d}\sin kd & \frac{H}{d}\sin ld & -\omega
\end{pmatrix}.
$$

Ex 8.8: The coefficient matrix for A, B, and C is

$$
\begin{pmatrix}
\iota\omega & f & \frac{g}{2d}[K_-^+][L_+^+] \\
f & -\iota\omega & -\frac{g}{2d}[K_+^+][L_-^+] \\
\frac{H}{2d}[K_-^-][L_+^-] & \frac{H}{2d}[K_+^-][L_-^-] & -\iota\omega
\end{pmatrix}.
$$

See (A.5).

Ex 8.9: The coefficient matrix for A, B, and C is

$$\begin{pmatrix} \iota\omega & \frac{f}{4}[K_+^+][L_+^-] & \frac{g}{d}[K_-^+] \\ \frac{f}{4}[K_+^-][L_+^+] & -\iota\omega & -\frac{g}{d}[L_-^+] \\ \frac{H}{d}[K_-^-] & \frac{H}{d}[L_-^-] & -\iota\omega \end{pmatrix}.$$

See (A.5).

Ex 8.10: The coefficient matrix for A, B, and C is

$$\begin{pmatrix} \iota\omega & \frac{f}{4}[K_+^-][L_-^+] & -\frac{g}{4d}[K_+^+][K_-^-][L_+^+] \\ \frac{f}{4}[K_+^+][L_+^-] & -\iota\omega & \frac{g}{4d}[K_+^+][L_+^+][L_-^-] \\ \frac{H}{4d}[K_+^+][K_-^-][L_+^-] & \frac{H}{4d}[L_+^+][L_-^-][K_+^-] & -\iota\omega \end{pmatrix}.$$

See (A.5).

Ex 8.11: The coefficient matrix for A, B, and C is

$$\begin{pmatrix} (\lambda^2 - 1) & 0 & \frac{2\iota g\Delta t}{d}\lambda \sin kd \\ 0 & (\lambda^2 - 1) & \frac{2\iota g\Delta t}{d}\lambda \sin ld \\ \frac{2\iota H\Delta t}{d}\lambda \sin kd & \frac{2\iota H\Delta t}{d}\lambda \sin ld & (\lambda^2 - 1) \end{pmatrix}.$$

Setting the determinant equal to zero, the resulting equation can be put in the form of (8.88), where

$$\sigma^2 = \frac{4c^2\Delta t^2}{d^2}\left(\sin^2 kd + \sin^2 ld\right).$$

Ex 8.12: The coefficient matrix for A, B, and C is

$$\begin{pmatrix} (\lambda^2 - 1) & 0 & -\frac{g\Delta t}{d}[K_-^+][L_+^+]\lambda \\ 0 & (\lambda^2 - 1) & -\frac{g\Delta t}{d}[K_+^+][L_+^+]\lambda \\ \frac{H\Delta t}{d}[K_-^-][L_+^-]\lambda & \frac{H\Delta t}{d}[K_+^-][L_-^-]\lambda & (\lambda^2 - 1) \end{pmatrix}.$$

See (A.5). Setting the determinant equal to zero, the resulting equation can be put in the form of (8.88), where

$$\sigma^2 = \frac{8c^2\Delta t^2}{d^2}\left(1 - \cos kd \cos ld\right).$$

Ex 8.13: The coefficient matrix for A, B, and C is

$$\begin{pmatrix} (\lambda^2 - 1) & 0 & -\frac{2g\Delta t}{d}[K_-^+]\lambda \\ 0 & (\lambda^2 - 1) & -\frac{2g\Delta t}{d}[L_-^+]\lambda \\ \frac{2H\Delta t}{d}[K_-^-]\lambda & \frac{2H\Delta t}{d}[L_-^-]\lambda & (\lambda^2 - 1) \end{pmatrix}.$$

See (A.5). Setting the determinant equal to zero, the resulting equation can be put in the form of (8.88), where

$$\sigma^2 = \frac{8c^2\Delta t^2}{d^2}\left(2 - \cos kd - \cos ld\right).$$

Ex 8.14: The coefficient matrix for A, B, and C is

$$\begin{pmatrix} \iota\omega & f & \frac{g}{d'}\left(1-e^{K'}\right) \\ f & -\iota\omega & \frac{g}{d'}Ye^{K'/2} \\ \frac{H}{d'}\left(1-e^{-K'}\right) & \frac{H}{d'}Ye^{-K'/2} & -\iota\omega \end{pmatrix},$$

where $K' = \iota kd'$, $L' = \iota ld'$, and $Y = \left(e^{L'/2} - e^{-L'/2}\right)$.

Ex 8.15: The coefficient matrix for A, B, and C is

$$\begin{pmatrix} \left(\lambda^3 - \lambda^2\right) & 0 & -\frac{g\Delta t}{12d}[K_-^+][L_+^+]\Lambda \\ 0 & \left(\lambda^3 - \lambda^2\right) & -\frac{g\Delta t}{12d}[K_+^+][L_-^+]\Lambda \\ \frac{H\Delta t}{12d}[K_-^-][L_+^-]\Lambda & \frac{H\Delta t}{12d}[K_+^-][L_-^-]\Lambda & \left(\lambda^3 - \lambda^2\right) \end{pmatrix},$$

where $\Lambda = 23\lambda^2 - 16\lambda + 5$. See (A.5). Setting the determinant equal to zero results in

$$\left(\lambda^3 - \lambda^2\right)^2 + \sigma^2\Lambda^2 = 0 \qquad (A.6)$$

where

$$\sigma^2 = \frac{c^2\Delta t^2}{18d^2}\left(1 - \cos kd \cos ld\right).$$

Equation (A.6) can be rewritten as

$$\lambda^3 - \lambda^2 = \pm i\sigma\Lambda = \pm\iota\sigma\left(23\lambda^2 - 16\lambda + 5\right),$$

which results in the two third-order polynomials desired. Although neither polynomial can be directly evaluated, they can be plotted and analyzed to show that $|\lambda| \leq 1$ only when $|\sigma| \leq 0.06$.

Ex 8.16: The coefficient matrix for A, B, and C is

$$\begin{pmatrix} \left(\lambda^3 - \lambda^2\right) & 0 & -\frac{g\Delta t}{12d}[K_-^+]\Lambda \\ 0 & \left(\lambda^3 - \lambda^2\right) & -\frac{g\Delta t}{12d}[L_-^+]\Lambda \\ \frac{H\Delta t}{12d}[K_-^-]\Lambda & \frac{H\Delta t}{12d}[L_-^-]\Lambda & \left(\lambda^3 - \lambda^2\right) \end{pmatrix},$$

where $\Lambda = 23\lambda^2 - 16\lambda + 5$. See (A.5). Setting the determinant equal to zero results in (A.6), where

$$\sigma^2 = \frac{c^2\Delta t^2}{72d^2}\left(2 - \cos kd - \cos ld\right).$$

Additional Time-differencing Schemes

B.1 Additional Time-differencing Schemes

In preceding chapters we have dealt extensively with four finite time-differencing schemes for equations of the form

$$\frac{\partial u}{\partial t} = F(u, \{\cdots\}), \tag{B.1}$$

where the forcing term $F(u, \{\cdots\})$ may be a function of one or more dependent variables. These additional variables are denoted by the bracketed ellipses, $\{\cdots\}$. The forcing term may also consist of multiple individual terms. The four schemes we have discussed are:

- The *forward scheme*, $u^{n+1} = u^n + \Delta t F(u^n)$, which is explicit.

- The *backward scheme*, $u^{n+1} = u^n + \Delta t F(u^{n+1})$, which is implicit.

- The *centered (or leapfrog) scheme*, $u^{n+1} = u^{n-1} + 2\Delta t F(u^n)$, which is explicit, and also has a computational mode (as does any multi-level scheme).

- The *3rd-order Adams-Bashforth scheme*, $u^{n+1} = u^n + (\Delta t/12)[23F(u^n) - 16F(u^{n-1}) + 5F(u^{n-2})]$, which is explicit and has two computational modes.

The specific stability criteria for these time-differencing schemes are dependent on the nature of the partial differential equations to which they are applied, as well as on the spatial differencing scheme that is employed.

We have focused on the four schemes above because they are widely used, not overly complicated, and readily illustrate the fundamental aspects of modeling. They are not the only possible time differencing schemes available, and in this section we present some additional schemes which are used in atmospheric modeling. However, our presentation of these schemes will not be as extensive as for the previous four schemes. We do not present detailed stability analyses, and only briefly discuss important features for each scheme.

B.2   Trapezoidal Scheme

The *trapezoidal scheme* (also called the *Crank-Nicolson scheme*) is an implicit scheme that uses the average of the forcing function at two time levels. It has the form [1]

$$u^{n+1} = u^n + \frac{\Delta t}{2}[F(u^n) + F(u^{n+1})]. \tag{B.2}$$

This scheme is always stable for both the advection and diffusion equations.

B.3   Predictor-corrector Schemes

One method of achieving greater accuracy is to predict an intermediate value of the variable u at some fractional time level between time levels n and $n + 1$, and then use this intermediate value to aid in predicting the value at $n + 1$. Such schemes are called *predictor-corrector*, or *multistage*, schemes. The number of stages is determined by the number of intermediate values that are used. Theoretically, the greater number of stages, the closer the convergence will be between the simulated and actual values of u^{n+1}.

---

[1]Some may consider the trapezoidal scheme to be semi-implicit, since we are evaluating the forcing term at both time levels n and $n + 1$. However, a true semi-implicit scheme would evaluate certain variables within the forcing term at time level n while other, different variables within the same forcing function at time level $n + 1$. This distinction is rather academic.

A general two-stage predictor corrector scheme has a *predictor step* given by

$$\tilde{u}^{n+\alpha} = u^{n-\beta} + (\alpha + \beta)\,\Delta t F\,(u^n)\,, \tag{B.3}$$

followed by a *corrector step* of the form

$$u^{n+1} = u^n + \Delta t[(1 - \gamma)F(u^n) + \gamma F(\tilde{u}^{n+\alpha})], \tag{B.4}$$

which uses a linear combination of the forcing terms at the time levels n and $n + \alpha$. The various predictor-corrector schemes differ in the choice of values for the parameters α, β, and γ.

B.3.1 Leapfrog-trapezoidal scheme

One possible predictor-corrector scheme is to use a leapfrog scheme to predict an intermediate value of u at time level $n + 1$,

$$\tilde{u}^{n+1} = u^{n-1} + 2\Delta t F(u^n). \tag{B.5}$$

This intermediate value is then used to estimate the forcing term at time level $n + 1$, which is then averaged with the forcing term at time n and used in a trapezoidal manner to compute the final value of u^{n+1},

$$u^{n+1} = u^n + \frac{\Delta t}{2}[F(u^n) + F(\tilde{u}^{n+1})]. \tag{B.6}$$

This is equivalent to (B.3) and (B.4) with $\alpha = \beta = 1$, and $\gamma = 1/2$. Although this scheme may appear to be implicit, it is actually explicit because it could be written in the form

$$u^{n+1} = u^n + \frac{\Delta t}{2}\{F(u^n) + F[u^{n-1} + 2\Delta t F(u^n)]\}. \tag{B.7}$$

B.3.2 The Matsuno scheme

The *Matsuno scheme* consists of a forward-in-time predictor step,

$$\tilde{u}^{n+1} = u^n + \Delta t F(u^n), \tag{B.8}$$

followed by a backward-in-time corrector step,

$$u^{n+1} = u^n + \Delta t F(\tilde{u}^{n+1}). \tag{B.9}$$

It is also called the *forward-backward scheme*, and is equivalent to (B.3) and (B.4) with $\alpha = 1$, $\beta = 0$, and $\gamma = 1$. The Matsuno scheme is highly damped and is not a good choice for most modeling applications. However, it has been successfully employed during the initialization phase of primitive equation models where it is periodically substituted in place of the leapfrog scheme in order to dampen high-frequency gravity waves and the computational mode.

B.3.3   Runge-Kutta schemes

Two-stage predictor-corrector schemes for which $\beta = 0$ and for which the product $\alpha\gamma = 1/2$ are called *Runge-Kutta* schemes. The case where $\alpha = 1$ and $\gamma = 1/2$ is called the *Heun* scheme, and is represented by the predictor and corrector steps

$$\tilde{u}^{n+1} = u^n + \Delta t F(u^n) \tag{B.10}$$

and

$$u^{n+1} = u^n + \frac{\Delta t}{2}[F(u^n) + F(\tilde{u}^{n+1})]. \tag{B.11}$$

The Heun scheme is nearly the same as the leapfrog-trapezoidal scheme, only the predictor step uses a forward difference rather than a leapfrog difference.

     The two-stage Runge-Kutta schemes are not particularly useful compared to other schemes, because of stability issues. However, there are higher-order Runge-Kutta schemes which, similar to the 3rd-order Adams-Bashforth scheme, have good stability and whose computational modes are highly damped.

     The 3rd-order Runge-Kutta scheme uses two intermediate steps,

$$\tilde{u}_1 = u^n + \frac{\Delta t}{3}F(u^n) \tag{B.12}$$

$$\tilde{u}_2 = \tilde{u}_1 + \frac{15\Delta t}{16}\left[F(\tilde{u}_1) - \frac{5}{9}F(u^n)\right], \tag{B.13}$$

followed by the final step

$$u^{n+1} = \tilde{u}_2 + \frac{8\Delta t}{15}\left(F(\tilde{u}_2) - \frac{153}{128}\left[F(\tilde{u}_1) - \frac{5}{9}F(u^n)\right]\right). \tag{B.14}$$

The 4rd-order Runge-Kutta scheme uses three intermediate steps

$$\tilde{u}_1 = u^n + \frac{\Delta t}{2} F(u^n) \tag{B.15}$$

$$\tilde{u}_2 = u^n + \frac{\Delta t}{2} F(\tilde{u}_1) \tag{B.16}$$

$$\tilde{u}_3 = u^n + \Delta t F(\tilde{u}_2), \tag{B.17}$$

prior to the final step

$$u^{n+1} = u^n + \frac{\Delta t}{6} \left[F(u^n) + 2F(\tilde{u}_1) + 2F(\tilde{u}_2) + F(\tilde{u}_3) \right]. \tag{B.18}$$

APPENDIX C

The Fourier Transform

C.1   The Fourier Transform

Fourier analysis is a mathematical tool that allows most functions to be written as a sum of sinusoids of different wave numbers[1] and amplitudes. The amplitudes of the sinusoids are known as the *Fourier* or *spectral coefficients*. If the Fourier coefficients are known, then the original function can be reconstructed from them. For continuous, periodic functions the form of Fourier analysis that is used is the *Fourier series*. For discrete data we instead use the *discrete Fourier transform*, or *DFT*. A commonly used computer algorithms for computing discrete Fourier transforms is the *fast Fourier transform* or *FFT*.[2]

---

[1]For spatial data, Fourier analysis decomposes the function into sinusoids of varying wave number. For temporal data, the decomposition is in terms of frequency. The underlying mathematics is the same in either case.

[2]The terms FFT and DFT are often used interchangeably; however, the term FFT really only refers to the fast Fourier transform algorithm for computing a DFT.

C.1.1   The DFT equations

The equations for the discrete Fourier transform are

$$U_m = \frac{1}{N} \sum_{i=0}^{N-1} u_i \exp\left(-\iota 2\pi im\right) \tag{C.1}$$

$$u_i = \sum_{m=0}^{N-1} U_m \exp\left(\iota 2\pi im\right). \tag{C.2}$$

Equation (C.1) is called the *forward transform*, and returns the spectral coefficients, U_m, from the input function, u_i. Equation (C.2) is called the *inverse transform*, *reverse transform*, or *backward transform*, and returns the original function from the spectral coefficients. In these equations i is the grid point number (beginning with zero), m is the spectral coefficient index, and N is the total number of data points. Notice that the number of spectral coefficients returned by the FFT is equal to the number of data points. The wave numbers of the spectral coefficients are given by

$$k_m = \frac{2\pi m}{Nd}; \quad m = 0, 1, 2, \cdots, N - 1, \tag{C.3}$$

where d is the distance between grid points. Due to symmetry properties of the DFT, the wave numbers for $m > N/2$ can also be thought of as being negative wave numbers, using the following identity

$$k_{N/2+b} = -k_{N/2-b} \tag{C.4}$$

for any integer $0 \le b \le N/2 - 1$. The wave numbers are then given by

$$k_m = \begin{cases} 2\pi m/Nd, & \text{for } 0 \le m \le N/2 \\ 2\pi\left(m - N\right)/Nd, & \text{for } N/2 + 1 \le m \le N - 1. \end{cases} \tag{C.5}$$

Most software packages return negative wave numbers using (C.5) when computing DFTs.

C.1.2   Alternate forms of the DFT equations

Equations (C.1) and (C.2) are not the only possible definitions for the DFT equations. Some sources define the forward and inverse transforms as

$$U_m = \sum_{i=0}^{N-1} u_i \exp\left(-\imath 2\pi i m\right) \qquad \text{(C.6)}$$

$$u_i = \frac{1}{N} \sum_{m=0}^{N-1} U_m \exp\left(\imath 2\pi i m\right), \qquad \text{(C.7)}$$

with the normalization factor, $1/N$, applied to the inverse transform rather than to the forward transform. Others split the normalization factor between the forward and inverse transform, defining them as

$$U_m = \frac{1}{\sqrt{N}} \sum_{i=0}^{N-1} u_i \exp\left(-\imath 2\pi i m\right) \qquad \text{(C.8)}$$

$$u_i = \frac{1}{\sqrt{N}} \sum_{m=0}^{N-1} U_m \exp\left(\imath 2\pi i m\right). \qquad \text{(C.9)}$$

When using a software package or library to perform DFTs, it is important to know which definitions are being used for the forward and inverse transforms.

C.2   A DFT Example

We illustrate the DFT by using a grid consisting of 101 grid points with indices ranging from $i = 0$ to 100, and a grid spacing of 1 meter. The values of u on this grid are generated using the equation

$$
\begin{aligned}
u_i = 1.5 &+ \cos(5k_0 i) + 4\cos(15k_0 i) \\
&- 3\cos(30k_0 i) + 2.5\cos(40k_0 i) + 2\sin(8k_0 i) \\
&+ 3\sin(15k_0 i) - 5\sin(35k_0 i), \quad \text{(C.10)}
\end{aligned}
$$

where $k_0 = 2\pi/(N-1)$. A plot of u_i is shown in the top panel of Fig. C-1.

When we take the DFT of u_i it results in 101 spectral coefficients, U_m. Even though the input function, u_i, is real-valued, the spectral coefficients are complex-valued. Plots of the real and imaginary components of the spectral coefficients as functions of wave number are shown in the middle and bottom panels of Fig. C2. The real

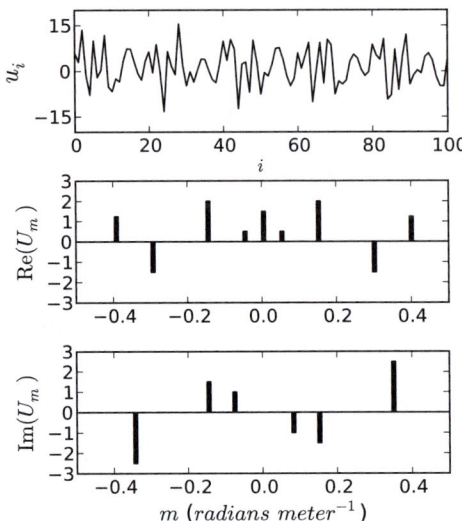

Fig. C.1: The top panel shows a plot of u_i from (C.10). The middle panel shows the real part of the Fourier coefficients, which correspond to the cosine terms. The bottom panel shows the imaginary part of the Fourier coefficients, which correspond to the sine terms. Note that the negative wave numbers are redundant, and could be ignored with no loss of information.

part of the spectral coefficients (middle panel) corresponds to the cosine terms in the original function, while the imaginary part (bottom panel) corresponds to the sine terms.

C.3 Real versus Complex DFTs

For real-valued input the Fourier coefficients for the negative wave numbers are the complex-conjugates of those for the positive wave numbers and do not contain any additional information. It is therefore redundant to compute the coefficients for all the wave numbers if the input function is real. Instead, we can just compute the coefficients for the zero and positive wave numbers. Many software applications have a *real DFT* function for doing this, which cuts the computational time and memory usage in half compared to using the complex DFT. It is important to remember that the coefficients themselves will still be complex-valued even if the input is real-valued.

Operator Splitting

D.1 Split versus Unsplit Schemes

Consider an equation of the form

$$\frac{\partial s}{\partial t} = F(s) + G(s).$$ (D.1)

This could represent the two-dimensional advection equation, in which case the operators F and G are given by

$$F = -u\frac{\partial}{\partial x}$$ (D.2)

and

$$G = -v\frac{\partial}{\partial y}.$$ (D.3)

Equation (D.1) might also represent the one-dimensional advection-diffusion equation, in which case the operators F and G would be

$$F = -u\frac{\partial}{\partial x}$$ (D.4)

and

$$G = K\frac{\partial^2}{\partial x^2}.$$ (D.5)

Using forward time differencing the predictive equation for s is

$$s^{n+1} = s^n + \Delta t F(s) + \Delta t G(s). \tag{D.6}$$

If instead we choose to use operator splitting, we would use the following two-step process,

$$\tilde{s} = s^n + \Delta t F(s^n) \tag{D.7}$$

$$s^{n+1} = \tilde{s} + \Delta t G(\tilde{s}). \tag{D.8}$$

The unsplit scheme, (D.6), and the split scheme, (D.7) and (D.8), will not necessarily yield the same answer. This can be seen by eliminating $\tilde{s}$ from (D.7) and (D.8) to get

$$s^{n+1} = s^n + \Delta t F(s^n) + \Delta t G(s^n) + \Delta t^2 G\left[F(s^n)\right]. \tag{D.9}$$

Equation (D.9) shows that the two schemes differ by the term $\Delta t^2 G[F(s^n)]$, which is called the *splitting term* or *splitting error*.

  We could have also reversed the order in which we applied the operators F and G by instead using the split scheme

$$\tilde{s} = s^n + \Delta t G(s^n) \tag{D.10}$$

$$s^{n+1} = \tilde{s} + \Delta t F(\tilde{s}), \tag{D.11}$$

which when combined yields

$$s^{n+1} = s^n + \Delta t F(s^n) + \Delta t G(s^n) + \Delta t^2 F\left[G(s^n)\right]. \tag{D.12}$$

Unless the operators F and G commute, (D.9) and (D.12) will not result in the same value for s^{n+1}.

  Both of the split schemes we have shown are biased, because the result depends on which operator, F or G, is applied first. To overcome this bias we could alternate the schemes at each time step, applying F first at all the even time steps, and G first at all the odd time steps. Or we could use both schemes at each time step and average them.

D.2   Why Use a Split Scheme?

  There are three main reasons why split schemes are used. The first is that a certain scheme may be stable for one of the terms in an equation, but unstable for the other. An example of this is the advection-diffusion equation, in which leapfrog time differencing cannot be applied to the diffusion term. In this case, if (D.1) represented the advection-diffusion equation with F being the advection operator (D.4) and G being the diffusion operator (D.5), then we could use the split scheme

$$\tilde{s} = s^{n-1} + 2\Delta t F(s^n) \tag{D.13}$$

$$s^{n+1} = \tilde{s} + \Delta t G(\tilde{s}). \tag{D.14}$$

The second reason for using a split scheme is that it may reduce the number of computations required. As an example, consider the application of the trapezoidal scheme[1] to the two-dimensional advection equation,

$$s^{n+1} = s^n + \frac{1}{2}[F(s^n) + F(s^{n+1})] + \frac{1}{2}[G(s^n) + G(s^{n+1})]. \tag{D.15}$$

This scheme is implicit, and for an $NX \times NY$ grid requires the solution of a system of $M = NX \times NY$ simultaneous equations. The coefficient matrix for this system of equations would be an $M \times M$ matrix. If instead we used the split procedure

$$\tilde{s} = s^n + \frac{1}{2}[F(s^n) + F(s^{n+1})] \tag{D.16}$$

$$s^{n+1} = \tilde{s} + \frac{1}{2}[G(\tilde{s}) + G(s^{n+1})], \tag{D.17}$$

we end up needing to implicitly solve two equations, each having M simultaneous equations. At first it would seem that the split system is at least twice as computationally intensive as the unsplit scheme, but it turns out otherwise. This is because the coefficient matrices for (D.16) and (D.17) are of a certain form called *tridiagonal*, meaning that the only nonzero elements are along the main (upper-left to lower-right) diagonal and the minor diagonal on either side of this main diagonal. Tridiagonal matrices are much easier to reduce and solve than are non-tridiagonal matrices. So, even though the split scheme requires the solution of two matrices, rather than a single matrix, it actually requires fewer computational steps than the unsplit scheme.

---

[1]See Appendix B for the definition of the trapezoidal scheme.

A third reason for using a split scheme is that certain schemes, such as the flux-form semi-Lagrangian advection schemes discussed in Chapter 15, involve complicated algorithms that include *if-then* constructs. These algorithms can only be applied in one dimension, and so if two dimensions are present the algorithm must be applied sequentially in a split mode.

D.3 Accounting for Splitting Error

The result of a split scheme will differ from that of an unsplit scheme by terms of the form $\Delta t^2 G[F(s^n)]$. To enhance the accuracy of the split scheme we could compute these terms and then subtract them from the split-scheme result. Depending on how large these splitting-error terms are, it may be desirable to account for them in this way. For flux-form advection calculations in nonhydrostatic compressible flows, the splitting-error correction terms can often be neglected. For simulations of large-scale hydrostatic flows they should be included.[2]

---

[2]Lin, S.-J. and R. B Rood, 1996: Multidimensional flux-form semi-Lagrangian transport schemes, *Mon. Wea. Rev.*, **124**, 2046-2070.

E.1   Vector Equations as Coordinate Invariants

The momentum equation written in vector form, and omitting the turbulent flux term, is

$$\frac{\partial \vec{V}}{\partial t} + \vec{V} \cdot \nabla \vec{V} = -\frac{1}{\rho} \nabla p - 2\vec{\Omega} \times \vec{V} + \vec{g}. \tag{E.1}$$

Although it appears that this is a single equation, it really is three scalar equations for $\partial u/\partial t$, $\partial v/\partial t$, and $\partial w/\partial t$. Equation (E.1) is *geometrically invariant*, meaning that it is valid in any coordinate system (e.g., spherical, cylindrical, Cartesian). Only when we decide on a particular coordinate geometry and expand the vectors into their components do the resulting equations differ between coordinate systems.

E.2   The Del Operator

The del operator expressed in Cartesian coordinates is

$$\nabla = \hat{i}\frac{\partial}{\partial x} + \hat{j}\frac{\partial}{\partial y} + \hat{k}\frac{\partial}{\partial z} \tag{E.2}$$

where $\hat{i}$, $\hat{j}$, and $\hat{k}$ are the unit vectors in the x, y, and z directions. In spherical coordinates, using r for the radial distance from the center of the Earth, ϕ for latitude, and λ for longitude, the del operator is expressed as

$$\nabla = \hat{i}' \frac{1}{r \cos \phi} \frac{\partial}{\partial \lambda} + \hat{j}' \frac{1}{r} \frac{\partial}{\partial \phi} + \hat{k}' \frac{\partial}{\partial r} \qquad (E.3)$$

where $\hat{i}'$, $\hat{j}'$, and $\hat{k}'$ are the unit vectors in the λ, ϕ, and r directions. The del operator is also geometrically invariant. A change of coordinate systems does not change the del operator itself, only its component representation.

E.2.1 Gradient

When the del operator is applied to a scalar, the result is a vector called the *gradient*. For an arbitrary scalar s the gradient in Cartesian coordinates is

$$\nabla s = \hat{i} \frac{\partial s}{\partial x} + \hat{j} \frac{\partial s}{\partial y} + \hat{k} \frac{\partial s}{\partial z}, \qquad (E.4)$$

while in spherical coordinates it is

$$\nabla s = \hat{i}' \frac{1}{r \cos \phi} \frac{\partial s}{\partial \lambda} + \hat{j}' \frac{1}{r} \frac{\partial s}{\partial \phi} + \hat{k}' \frac{\partial s}{\partial r}. \qquad (E.5)$$

E.2.2 Divergence

Dotting the del operator with a vector results in the *divergence* of the vector field. In Cartesian coordinates the component-form of vector $\vec{A}$ is

$$\vec{A} = A_x \hat{i} + A_y \hat{j} + A_z \hat{k}, \qquad (E.6)$$

and the divergence is

$$(\hat{i} \frac{\partial}{\partial x} + \hat{j} \frac{\partial}{\partial y} + \hat{k} \frac{\partial}{\partial z}) \cdot (A_x \hat{i} + A_y \hat{j} + A_z \hat{k}) =$$

$$\hat{i} \cdot \frac{\partial}{\partial x}(A_x \hat{i} + A_y \hat{j} + A_z \hat{k}) + \hat{j} \cdot \frac{\partial}{\partial y}(A_x \hat{i} + A_y \hat{j} + A_z \hat{k})$$

$$+ \hat{k} \cdot \frac{\partial}{\partial z}(A_x \hat{i} + A_y \hat{j} + A_z \hat{k}). \qquad (E.7)$$

Since the unit vectors $\hat{i}$, $\hat{j}$, and $\hat{k}$ are constant in both magnitude and direction, any spatial derivative of the unit vectors are zero. Also, since the unit vectors are orthogonal, the dot product between distinct unit vectors $(\hat{i} \cdot \hat{j}, \hat{i} \cdot \hat{k}, \hat{j} \cdot \hat{k})$ is zero, while that between identical unit vectors $(\hat{i} \cdot \hat{i}, \hat{j} \cdot \hat{j}, \hat{k} \cdot \hat{k})$ is one. Therefore, the result is the familiar

$$\nabla \cdot \vec{A} = \frac{\partial A_x}{\partial x} + \frac{\partial A_y}{\partial y} + \frac{\partial A_z}{\partial z}. \tag{E.8}$$

In spherical coordinates the dot product is more complicated, since the unit vectors $\hat{i}'$, $\hat{j}'$, and $\hat{k}'$, though orthogonal, are not constant in direction. In spherical coordinates the component-form of vector $\vec{A}$ is

$$\vec{A} = A_\lambda \hat{i}' + A_\phi \hat{j}' + A_r \hat{k}', \tag{E.9}$$

and the divergence is

$$(\hat{i}' \frac{1}{r \cos \phi} \frac{\partial}{\partial \lambda} + \hat{j}' \frac{1}{r} \frac{\partial}{\partial \phi} + \hat{k}' \frac{\partial}{\partial r}) \cdot (A_\lambda \hat{i}' + A_\phi \hat{j}' + A_r \hat{k}') =$$

$$\frac{1}{r \cos \phi} \hat{i}' \cdot \frac{\partial}{\partial \lambda} \cdot (A_\lambda \hat{i}' + A_\phi \hat{j}' + A_r \hat{k}')$$

$$+ \frac{1}{r} \hat{j}' \cdot \frac{\partial}{\partial \phi} \cdot (A_\lambda \hat{i}' + A_\phi \hat{j}' + A_r \hat{k}')$$

$$+ \hat{k}' \cdot \frac{\partial}{\partial r} \cdot (A_\lambda \hat{i}' + A_\phi \hat{j}' + A_r \hat{k}'). \tag{E.10}$$

When expanded, (E.10) becomes

$$\nabla \cdot \vec{A} = \frac{1}{r \cos \phi} \frac{\partial A_\lambda}{\partial \lambda} + \frac{1}{r} \frac{\partial A_\phi}{\partial \phi} + \frac{\partial A_r}{\partial r}$$

$$+ \frac{1}{r \cos \phi} \hat{i}' \cdot (A_\lambda \frac{\partial \hat{i}'}{\partial \lambda} + A_\phi \frac{\partial \hat{j}'}{\partial \lambda} + A_r \frac{\partial \hat{k}'}{\partial \lambda})$$

$$+ \frac{1}{r} \hat{j}' \cdot (A_\lambda \frac{\partial \hat{i}'}{\partial \phi} + A_\phi \frac{\partial \hat{j}'}{\partial \phi} + A_r \frac{\partial \hat{k}'}{\partial \phi})$$

$$+ \hat{k}' \cdot (A_\lambda \frac{\partial \hat{i}'}{\partial r} + A_\phi \frac{\partial \hat{j}'}{\partial r} + A_r \frac{\partial \hat{k}'}{\partial r}). \tag{E.11}$$

The spatial derivatives of the unit vectors are given by

$$\frac{\partial \hat{i}'}{\partial \lambda} = \hat{j}' \sin \phi - \hat{k}' \cos \phi, \frac{\partial \hat{i}'}{\partial \phi} = 0, \frac{\partial \hat{i}'}{\partial r} = 0,$$

$$\frac{\partial \hat{j}'}{\partial \lambda} = -\hat{i}' \sin \phi, \frac{\partial \hat{j}'}{\partial \phi} = -\hat{k}', \frac{\partial \hat{j}'}{\partial r} = 0, \qquad (E.12)$$

$$\frac{\partial \hat{k}'}{\partial \lambda} = -\hat{i}' \cos \phi, \frac{\partial \hat{k}'}{\partial \phi} = \hat{j}', \frac{\partial \hat{k}'}{\partial r} = 0,$$

and when these are substituted into (E.11) the result for the divergence in spherical coordinates is[1]

$$\nabla \cdot \vec{A} = \frac{1}{r \cos \phi} \frac{\partial A_\lambda}{\partial \lambda} + \frac{1}{r} \frac{\partial A_\phi}{\partial \phi} + \frac{\partial A_r}{\partial r} - \frac{A_\phi}{r} \tan \phi. \qquad (E.13)$$

E.3 Momentum Advection

In the advective term $\vec{V} \cdot \nabla \vec{V}$ of the momentum equation the del operator is applied to the velocity vector. Although this is a perfectly legitimate mathematical operation, it results in a cumbersome second-order tensor. For this reason it is convenient to use the identity[2]

$$\vec{V} \cdot \nabla \vec{V} = (\vec{V} \cdot \nabla) \vec{V}, \qquad (E.14)$$

where the quantity in parentheses is the *advection operator*. In Cartesian coordinates the advection operator expands to

$$\vec{V} \cdot \nabla = u \frac{\partial}{\partial x} + v \frac{\partial}{\partial y} + w \frac{\partial}{\partial z} \qquad (E.15)$$

while in spherical coordinates it expands to

$$\vec{V} \cdot \nabla = \frac{u}{r \cos \phi} \frac{\partial}{\partial \lambda} + \frac{v}{r} \frac{\partial}{\partial \phi} + w \frac{\partial}{\partial r}. \qquad (E.16)$$

---

[1]We thank Todd Sikora for providing a significant correction to the original draft of this section.

[2]See DeCaria, A. J. and T. D. Sikora, 2010: Momentum advection and the gradient of a vector field: A discussion of standard notation, *J. Atmos. Sci.*, **67**, 1287-1291.

The advective term in Cartesian coordinates is then

$$(\vec{V} \cdot \nabla)\vec{V} = u\frac{\partial}{\partial x}(u\hat{i} + v\hat{j} + w\hat{k})$$
$$+ v\frac{\partial}{\partial y}(u\hat{i} + v\hat{j} + w\hat{k}) + w\frac{\partial}{\partial z}(u\hat{i} + v\hat{j} + w\hat{k}), \quad (E.17)$$

while in spherical coordinates it is

$$(\vec{V} \cdot \nabla)\vec{V} = \frac{u}{r\cos\phi}\frac{\partial}{\partial\lambda}(u\hat{i'} + v\hat{j'} + w\hat{k'})$$
$$+ \frac{v}{r}\frac{\partial}{\partial\phi}(u\hat{i'} + v\hat{j'} + w\hat{k'}) + w\frac{\partial}{\partial r}(u\hat{i'} + v\hat{j'} + w\hat{k'}). \quad (E.18)$$

In Cartesian coordinates (E.17) expands to

$$(\vec{V} \cdot \nabla)\vec{V} = \left(u\frac{\partial u}{\partial x} + v\frac{\partial u}{\partial y} + w\frac{\partial u}{\partial z}\right)\hat{i}$$
$$+ \left(u\frac{\partial v}{\partial x} + v\frac{\partial v}{\partial y} + w\frac{\partial v}{\partial z}\right)\hat{j}$$
$$+ \left(u\frac{\partial w}{\partial x} + v\frac{\partial w}{\partial y} + w\frac{\partial w}{\partial z}\right)\hat{k}, \quad (E.19)$$

while in spherical coordinates (E.18) expands to

$$(\vec{V} \cdot \nabla)\vec{V} =$$
$$\frac{u}{r\cos\phi}\left(\frac{\partial u}{\partial\lambda}\hat{i'} + u\frac{\partial\hat{i'}}{\partial\lambda} + \frac{\partial v}{\partial\lambda}\hat{j'} + v\frac{\partial\hat{j'}}{\partial\lambda} + \frac{\partial w}{\partial\lambda}\hat{k'} + w\frac{\partial\hat{k'}}{\partial\lambda}\right)$$
$$+ \frac{v}{r}\left(\frac{\partial u}{\partial\phi}\hat{i'} + u\frac{\partial\hat{i'}}{\partial\phi} + \frac{\partial v}{\partial\phi}\hat{j'} + v\frac{\partial\hat{j'}}{\partial\phi} + \frac{\partial w}{\partial\phi}\hat{k'} + w\frac{\partial\hat{k'}}{\partial\phi}\right) \quad (E.20)$$
$$+ w\left(\frac{\partial u}{\partial r}\hat{i'} + u\frac{\partial\hat{i'}}{\partial r} + \frac{\partial v}{\partial r}\hat{j'} + v\frac{\partial\hat{j'}}{\partial r} + \frac{\partial w}{\partial r}\hat{k'} + w\frac{\partial\hat{k'}}{\partial r}\right).$$

When the spatial derivatives of the unit vectors, (E.12) are substituted into (E.20) the result is

$$(\vec{V} \cdot \nabla)\vec{V} =$$

$$\left(\frac{u}{r\cos\phi} \frac{\partial u}{\partial \lambda} + \frac{v}{r} \frac{\partial u}{\partial \phi} + w\frac{\partial u}{\partial r} - \frac{uv\tan\phi}{r} + \frac{uw}{r} \right) \hat{i}'$$

$$+ \left(\frac{u}{r\cos\phi} \frac{\partial v}{\partial \lambda} + \frac{v}{r} \frac{\partial v}{\partial \phi} + w\frac{\partial v}{\partial r} - \frac{u^2\tan\phi}{r} + \frac{vw}{r} \right) \hat{j}' \qquad \text{(E.21)}$$

$$+ \left(\frac{u}{r\cos\phi} \frac{\partial w}{\partial \lambda} + \frac{v}{r} \frac{\partial w}{\partial \phi} + w\frac{\partial w}{\partial r} - \frac{u^2+v^2}{r} \right) \hat{k}'.$$

E.4　The Momentum Equations

In spherical coordinates the components of the momentum equations (without turbulent flux terms) are

$$\frac{\partial u}{\partial t} + \frac{u}{r\cos\phi} \frac{\partial u}{\partial \lambda} + \frac{v}{r} \frac{\partial u}{\partial \phi} + w\frac{\partial u}{\partial r} - \frac{uv\tan\phi}{r} + \frac{uw}{r} =$$

$$-\frac{1}{\rho r\cos\phi} \frac{\partial p}{\partial \lambda} + 2\Omega\sin\phi\, v - 2\Omega\cos\phi\, w \quad \text{(E.22)}$$

$$\frac{\partial v}{\partial t} + \frac{u}{r\cos\phi} \frac{\partial v}{\partial \lambda} + \frac{v}{r} \frac{\partial v}{\partial \phi} + w\frac{\partial v}{\partial r} - \frac{u^2\tan\phi}{r} + \frac{vw}{r} =$$

$$-\frac{1}{\rho r} \frac{\partial p}{\partial \phi} - 2\Omega\sin\phi\, u \quad \text{(E.23)}$$

$$\frac{\partial v}{\partial t} + \frac{u}{r\cos\phi} \frac{\partial w}{\partial \lambda} + \frac{v}{r} \frac{\partial w}{\partial \phi} + w\frac{\partial w}{\partial r} - \frac{u^2+v^2}{r} =$$

$$-\frac{1}{\rho} \frac{\partial p}{\partial r} + 2\Omega\cos\phi\, u - g. \quad \text{(E.24)}$$

We are not constrained to using λ, ϕ, and r as our spatial variables in spherical coordinates. We could also use the parameterized independent variables x, y, and z using the identities

$$z = r - a \qquad \text{(E.25)}$$

$$dx = r\cos\phi\, d\lambda \qquad \text{(E.26)}$$

$$dy = r d\phi \qquad \text{(E.27)}$$

$$dz = dr, \qquad \text{(E.28)}$$

where a is the mean radius of the Earth. From (E.25) thru (E.27) we can establish the following relations between the spatial derivatives in λ, ϕ, and r, and those in x, y, and z,

$$\frac{\partial}{\partial\lambda} = \frac{\partial x}{\partial\lambda}\frac{\partial}{\partial x} = r\cos\phi\frac{\partial}{\partial x} \tag{E.29}$$

$$\frac{\partial}{\partial\phi} = \frac{\partial y}{\partial\phi}\frac{\partial}{\partial y} = r\frac{\partial}{\partial y} \tag{E.30}$$

$$\frac{\partial}{\partial r} = \frac{\partial z}{\partial r}\frac{\partial}{\partial z} = \frac{\partial}{\partial z}. \tag{E.31}$$

Using (E.29) thru (E.31) in (E.22) thru (E.25) results in

$$\frac{\partial u}{\partial t} + u\frac{\partial u}{\partial x} + v\frac{\partial u}{\partial y} + w\frac{\partial u}{\partial z} + \frac{uw - uv\tan\phi}{a+z} =$$
$$-\frac{1}{\rho}\frac{\partial p}{\partial x} + 2\Omega\sin\phi\,v - 2\Omega\cos\phi\,w \tag{E.32}$$

$$\frac{\partial v}{\partial t} + u\frac{\partial v}{\partial x} + v\frac{\partial v}{\partial y} + w\frac{\partial v}{\partial z} + \frac{vw + u^2\tan\phi}{a+z} =$$
$$-\frac{1}{\rho}\frac{\partial p}{\partial y} - 2\Omega\sin\phi\,u \tag{E.33}$$

$$\frac{\partial w}{\partial t} + u\frac{\partial w}{\partial x} + v\frac{\partial w}{\partial y} + w\frac{\partial w}{\partial z} + \frac{u^2 + v^2}{a+z} =$$
$$-\frac{1}{\rho}\frac{\partial p}{\partial z} + 2\Omega\cos\phi\,u - g. \tag{E.34}$$

Equations (E.32) thru (E.34) are written in spherical coordinates, even though the spatial variables are x, y, and z. The terms with the denominator $(a + z)$ are called the *curvature terms*, and would not appear in the Cartesian-coordinate form of the equations.

E.5 Curvature Terms in Other Equations

In spherical coordinates, curvature terms may appear in any equation where non-zero spatial derivatives of the unit vectors occur. Thus, any terms which contain the gradient of a vector $(\nabla\vec{A})$, the curl ($\nabla \times \vec{A}$) or divergence ($\nabla \cdot \vec{A}$) of a vector, or the Laplacian of a scalar or vector ($\nabla^2 s$ or $\nabla^2 \vec{A}$), are likely to result in curvature terms when written in component form in spherical coordinates. The gradient of a scalar (∇s) will not result in curvature terms.

As a further example, consider the continuity equation,

$$\frac{\partial \rho}{\partial t} + \nabla \cdot (\rho \vec{V}) = 0. \tag{E.35}$$

Expanded out in spherical coordinates and using (E.13), (E.35) is

$$\frac{\partial \rho}{\partial t} + \frac{1}{r \cos \phi} \frac{\partial}{\partial \lambda} (\rho u) + \frac{1}{r} \frac{\partial}{\partial \phi} (\rho v) + \frac{\partial}{\partial r} (\rho w) - \frac{\rho v}{r} \tan \phi = 0. \tag{E.36}$$

Put in terms of the parameterized variables x, y, and z using (E.29) thru (E.31), (E.36) becomes

$$\frac{\partial \rho}{\partial t} + \frac{\partial}{\partial x} (\rho u) + \frac{\partial}{\partial y} (\rho v) + \frac{\partial}{\partial z} (\rho w) - \frac{\rho v}{a + z} \tan \phi = 0. \tag{E.37}$$

Equation (E.37) is nearly identical to the flux-form continuity equation in Cartesian coordinates, with the exception of the additional curvature term.

APPENDIX F

Sources and Further Reading

The authors have found the following books, monographs, and articles to be particularly useful in preparing this text. The interested reader may wish to consult them for details and further reading.

Ahijevych, D., E. Gilleland, B. G. Brown, and E. E. Ebert, 2009: Application of spatial verification methods to idealized and NWP-gridded precipitation forecasts. *Wea. Forecasting*, **24**, 1485-1497.

Allen, D. J., A. R. Douglass, R. B. Rood and P. D. Guthrie, 1991: Application of a monotonic upstream-biased transport scheme to three-dimensional constituent transport calculations, *Mon. Wea. Rev.*, **119**, 2456-2464

Barth, A., A. A. Azcarate, P. Joassin, J.-M. Beckers, and C. Troupin, 2008: *Introduction to Optimal Interpolation and Variational Analysis*, GHER, 35 pp.

Bossavy, A., R. Girard and G. Kariniotakis, 2013: Forecasting ramps of wind power production with numerical weather prediction ensembles. *Wind Energy*, **16**, 51-63

Bouttier, F. and P. Courtier, 2002: *Data Assimilation Concepts and Methods*, ECMWF Training Course, 59 pp.

Chapman, D. C., 1985: Numerical treatment of cross-shelf open boundaries in a barotropic coastal ocean model, *J. Phys. Oceano.*, **15**, 1060-1075

Daley, R., 1991: *Atmospheric Data Analysis*, Cambridge University Press, 457 pp.

Durran, D. R., 1999: *Numerical Methods for Wave Equations in Geophysical Fluid Dynamics*, Springer, 465 pp.

Emanuel, K. and D. J. Raymond, 1993: *The Representation of Cumulus Convection in Numerical Models*, American Meteorological Society, 246 pp.

Haltiner, G. J. and R. T. Williams, 1980: *Numerical Weather Prediction*, Wiley, 477 pp.

Jacobson, M. Z., 2005: *Fundamentals of Atmospheric Modeling*, Cambridge University Press, 813 pp.

Kalnay, E., 2003: *Atmospheric Modeling, Data Assimilation and Predictability*, Cambridge University Press, 341 pp.

Kantha, L. H. and C. A. Clayson, 2000: *Numerical Models of Oceans and Oceanic Processes*, Academic Press, 940 pp.

Krishnamurti, T. N. and L. Bounoua, 1996: *An Introduction to Numerical Weather Prediction Techniques*, CRC Press, 293 pp.

Leung L. R., and S. J. Ghan, 1995: A subgrid parameterization of orographic precipitation, *Theor. and Appl. Climatol.*, **52**, 95-118

Lin, S-J and R. B. Rood, 1996: Multidimensional flux-form semi-Lagrangian transport schemes, *Mon. Wea. Rev.*, **124**, 2046-2070

Lynch, P., 2006: *The Emergence of Numerical Weather Prediction: Richardson's Dream*, Cambridge University Press, 279 pp.

Mesinger, F. and A. Arakawa, 1976: *Numerical Methods Used in Atmospheric Models, Vol. I*, WMO, 64 pp.

Mandel, J., L. S. Bennethum, J. D. Beezley, J. L. Coen, C. C. Douglas, Kim, M., and Vodacek A., 2008: A wildfire model with data assimilation, *Math. and Comp. in Simulation*, **79**, 584-606.

Mathiesen, P. and J. Kleissl, 2011: Evaluation of numerical weather prediction for intra-day solar forecasting in the continental United States, *Solar Energy*, **85**,967-977

McCollor, Doug, Roland Stull, 2008: Hydrometeorological accuracy enhancement via postprocessing of numerical weather forecasts in complex terrain, *Wea. Forecasting*, **23**, 131-144.

Murphy, A. H., 1988: Skill scores based on the mean square error and their relationships to the correlation coefficient, *Mon. Wea. Rev.*, **116**, 2417-2424

Murphy, A. H. and H. Daan, 1985: Forecast Evaluation, *Probability, Statistics and Decision Making in Atmospheric Science*, A. H. Murphy and R. W. Katz, Eds., Westview Press, 379-437.

Pielke, R. A., 1984: *Mesoscale Meteorological Modeling*, Academic Press, 612 pp.

Reichle, R. H., J. P. Walker, R. D. Koster, P. R. Houser, 2002: Extended versus ensemble Kalman filtering for land data assimilation. *J. Hydrometeor.*, **3**, 728-740.

Roache, P. J., 1998: *Fundamentals of Computational Fluid Dynamics*, Hermosa Publishers, 648 pp.

Skamarock, W. C. 2004: Evaluating mesoscale NWP models using kinetic energy spectra, *Mon. Wea. Rev.*, **132**, 319-332

Stensrud, D. J., 2007: *Parameterization Schemes: Keys to Understanding Numerical Weather Prediction Models*, Cambridge University Press. 480 pp.

Tang, Y. and R. Grimshaw, 1996: Radiation boundary conditions in barotropic coastal ocean numerical models, *Comp. Phys.*, **123**, 96-110

Thompson, P. D., 1961: *Numerical Weather Analysis and Prediction*, The MacMillan Company, 170 pp.

Tonttila1 J., P. Räisänen, and H. Järvinen, 2013: Monte Carlo-based subgrid parameterization of vertical velocity and stratiform cloud microphysics in ECHAM5.5-HAM22, *Atmos. Chem. Phys.*, **13**, 7551-7565

Torn, R. D., and G. J. Hakim, 2008: Ensemble-based sensitivity analysis, *Mon. Wea. Rev.*, **136**, 663-677.

UCAR MetEd: *Understanding Data Assimilation: How Models Create Their Initial Conditions*,
`https://www.meted.ucar.edu/training_module.php?id=62`

Van Knowe, G. E., K. T. Waight, M. Ceperuelo, J. Aymamí, S. Parés, S. Arumugam, and J. Oh , 2010: A comparison of statistical and explicit short-term hydrological forecasting techniques, AMS 24th Conference on Hydrology

Wernli, H., C. Hofmann, M. Zimmer, 2009: Spatial forecast verification methods intercomparison project: Application of the SAL technique, *Wea. Forecasting*, **24**, 1472-1484.

Wilks, D. S., 1995: *Statistical Methods in the Atmospheric Sciences, 2ed.*, Academic Press, 627 pp.

Index